Human Biology Step by Step

From DNA & Cells to Systems & Physiology: The Science of Being Human

Michael Roberts

© 2024 by Michael Roberts

All rights reserved.

No part of this publication may be reproduced, distributed, or transmitted in any form or by any means, including photocopying, recording, or other electronic or mechanical methods, without the prior written permission of the publisher, except in the case of brief quotations embodied in critical reviews and certain other noncommercial uses permitted by U.S. copyright law.

This book is intended to provide general information on the subjects covered and is presented with the understanding that the author and publisher are not providing professional advice or services. While every effort has been made to ensure the accuracy and completeness of the information contained herein, neither the author nor the publisher guarantees such accuracy or completeness, nor shall they be responsible for any errors or omissions or for the results obtained from the use of such information. The contents of this book are provided "as is" and without warranties of any kind, either express or implied.

Publisher email: info@tagvault.org

PREFACE

Welcome to *Human Biology Step by Step*. This book is a journey into the fascinating field of human biology—a field that not only helps us understand how our bodies work but also provides insights into what makes us, quite literally, human.

Human biology is the cornerstone of medicine, science, and many aspects of everyday life. Whether it's the beating of your heart, the air filling your lungs, or the complex symphony of neurons firing in your brain, biology is at the core of it all. Yet, for many, it can seem overwhelming—a labyrinth of technical terms, processes, and concepts. My goal in writing this book is to demystify the subject and make it approachable, understandable, and even enjoyable for readers of all backgrounds.

This book is designed to take you through the journey of human biology step by step. From the foundational building blocks of life to the complexity of entire systems, you'll gain a complete understanding of how the human body works. Whether you're a student, a curious reader, or a professional looking to refresh your knowledge, this book provides a clear and structured path through the complexities of biology.

In the first chapter, we'll lay the groundwork by exploring the history of human biology, the tools scientists use to study the body, and the key concept of homeostasis. This foundational knowledge will help you see the big picture before we go deeper into individual topics.

As you move through the chapters, you'll discover the inspiring mechanisms that keep us alive and functioning. You'll learn about the chemistry of life, the marvel of DNA and genetics, and the incredible structure and function of cells. From there, we'll journey through tissues, bones, muscles, and nerves, uncovering how these components work in harmony to create the human experience.

Each chapter focuses on a specific system, breaking down complex ideas into digestible, relatable sections. We'll explore how the cardiovascular system delivers life-sustaining nutrients to every cell, how the respiratory system exchanges vital gases, and how the digestive system transforms food into energy. You'll also learn how the kidneys maintain balance, how the immune system protects us, and how our endocrine system orchestrates countless processes through hormones.

In writing this book, I wanted to ensure we didn't just focus on how these systems function but also on their real-world significance. For example, how does the nervous system adapt to new experiences? How does our gut microbiome influence not just digestion but our overall health? How do advances in biotechnology and artificial organs impact the future of medicine? These questions aren't just fascinating. They're vital for understanding the role human biology has in our rapidly changing world.

Chapter 16 ties it all together, examining how these systems interconnect and depend on one another to maintain balance. The human body is not a collection of isolated parts—it's an extraordinary, integrated whole. This chapter also looks to the future, exploring emerging research and technologies that promise to reshape how we understand and care for the human body.

To make this book as useful as possible, I've included an appendix with terms and definitions. If you encounter unfamiliar words or concepts, you can quickly look them up and continue reading with confidence.

Whether you're reading this book for academic purposes, personal interest, or professional development, I hope you find it engaging and rewarding. Let this book be your guide as you uncover the science of being human, one step at a time.

TOPICAL OUTLINE

Chapter 1: Foundations of Human Biology
- The History of Human Biology: From Ancient Anatomy to Modern Science
- The Role of Human Biology in Medicine and Technology
- Tools and Techniques for Studying the Human Body
- The Importance of Homeostasis in Maintaining Life
- Human Biology in Evolutionary Context: From Primates to Homo sapiens
- The Human Microbiome: The Body's Hidden Ecosystem
- Ethics in Human Biology: Challenges and Considerations

Chapter 2: The Building Blocks of Life
- The Chemistry of Life: Atoms, Molecules, and Biomolecules
- Overview of Organic Molecules: Proteins, Carbohydrates, Lipids, and Nucleic Acids
- The Role of Water in Human Biology

Chapter 3: DNA and Genetics
- Structure and Function of DNA: The Blueprint of Life
- Gene Expression: From DNA to Protein
- Inheritance Patterns and Genetic Disorders
- Modern Genetics: CRISPR and the Future of Gene Editing

Chapter 4: The Cell - Life's Fundamental Unit
- Cellular Structures and Their Functions: Organelles and Membranes
- Cellular Respiration and Energy Production (ATP)
- The Cell Cycle and Mitosis
- Apoptosis: The Science of Programmed Cell Death

Chapter 5: Tissues and Their Functions
- Classification of Tissues: Epithelial, Connective, Muscle, and Nervous
- Specialized Functions of Each Tissue Type
- How Tissues Work Together to Form Organs
- Tissue Repair and Regeneration: The Role of Stem Cells

Chapter 6: The Skeletal System
- Structure and Function of Bones
- Bone Development and Growth: Ossification
- The Role of Joints and Cartilage
- Calcium Homeostasis and Bone Health

Chapter 7: The Muscular System
- Types of Muscles: Skeletal, Cardiac, and Smooth

- Muscle Contraction: The Role of Actin, Myosin, and Calcium
- Exercise and Muscle Health

Chapter 8: The Nervous System
- Anatomy of the Nervous System: Central and Peripheral Divisions
- How Neurons Communicate: Synapses and Neurotransmitters
- Reflexes and the Autonomic Nervous System
- Neural Plasticity: How the Brain Adapts and Changes

Chapter 9: The Endocrine System
- Glands and Hormones: Key Players in Regulation
- The Role of the Endocrine System in Growth, Metabolism, and Reproduction
- Feedback Mechanisms and Hormonal Balance
- Endocrine Disruptors and Their Impact on Health

Chapter 10: The Cardiovascular System
- Anatomy of the Heart and Blood Vessels
- The Circulatory Pathway: Systemic and Pulmonary Circulation
- Blood Composition and Its Functions
- The Lymphatic System: Its Role in Circulation and Immunity

Chapter 11: The Respiratory System
- Anatomy of the Respiratory System: From Nose to Alveoli
- Gas Exchange: Oxygen and Carbon Dioxide Transport
- The Role of the Respiratory System in Acid-Base Balance

Chapter 12: The Digestive System
- The Digestive Tract: Anatomy and Functions of Each Organ
- Enzymes and Nutrient Absorption
- The Role of Gut Flora in Digestion

Chapter 13: The Urinary System
- Anatomy and Function of the Kidneys
- How the Body Maintains Water and Electrolyte Balance
- The Process of Urine Formation
- Acid-Base Regulation in the Kidneys

Chapter 14: The Immune System
- Innate vs. Adaptive Immunity: First and Second Lines of Defense
- How Vaccines Work and Their Impact on Health
- Disorders of the Immune System: Autoimmunity and Allergies

Chapter 15: The Reproductive System
- Anatomy of Male and Female Reproductive Systems

- Hormonal Control of Reproduction
- The Biology of Fertilization and Pregnancy
- Reproductive Technologies: Innovations and Ethical Considerations

Chapter 16: Integration and Interdependence in Human Biology
- The Relationship Between Body Systems in Maintaining Homeostasis
- How Stress Affects Biological Systems
- Advances in Artificial Organs and Biotechnology
- Future Directions in Human Biology Research

Appendix
- Terms and Definitions

Afterword

TABLE OF CONTENTS

Chapter 1: Foundations of Human Biology ... 1
Chapter 2: The Building Blocks of Life ... 21
Chapter 3: DNA and Genetics ... 28
Chapter 4: The Cell - Life's Fundamental Unit ... 40
Chapter 5: Tissues and Their Functions ... 51
Chapter 6: The Skeletal System ... 62
Chapter 7: The Muscular System ... 73
Chapter 8: The Nervous System ... 79
Chapter 9: The Endocrine System ... 90
Chapter 10: The Cardiovascular System ... 100
Chapter 11: The Respiratory System ... 109
Chapter 12: The Digestive System ... 115
Chapter 13: The Urinary System ... 121
Chapter 14: The Immune System ... 128
Chapter 15: The Reproductive System ... 135
Chapter 16: Integration and Interdependence in Human Biology ... 142
Appendix ... 149
Afterword ... 152

CHAPTER 1: FOUNDATIONS OF HUMAN BIOLOGY

The History of Human Biology: From Ancient Anatomy to Modern Science

The story of human biology begins thousands of years ago, rooted in curiosity about the human body and its functions. Ancient civilizations sought to understand the body through observation, dissection, and experimentation. Each era contributed a foundation that shaped the sophisticated science we recognize today.

Ancient Beginnings: Anatomical Curiosity in Egypt and Mesopotamia

Ancient Egyptians were among the first to systematically study the human body. Their knowledge of anatomy came primarily from mummification. While their primary goal was preservation for the afterlife, the process involved removing organs like the liver, stomach, lungs, and intestines. This gave them a basic understanding of internal anatomy. They even documented their findings in medical texts such as the *Edwin Smith Papyrus* (circa 1600 BCE), which detailed surgical procedures and injuries. These records reveal an awareness of the brain, although its function remained a mystery at the time.

In Mesopotamia, early medical practices focused on symptom descriptions and treatments, blending empirical observations with spiritual interpretations. They did not perform dissections, as the body was considered sacred, but they noted physical signs such as changes in the pulse. These observations laid early groundwork for recognizing patterns in disease.

Greek Contributions: Dissection and the Rise of Rational Thought

The ancient Greeks propelled human biology forward by prioritizing reason over myth. Hippocrates (460–370 BCE), often called the "Father of Medicine," rejected supernatural explanations for illness, advocating for natural causes instead. He emphasized the balance of the four humors—blood, phlegm, yellow bile, and black bile—as critical to health. Though this theory was incorrect, it dominated medical thought for over a millennium.

The Greek anatomist Herophilos (335–280 BCE) conducted some of the earliest known dissections on human cadavers. Working in Alexandria, Egypt, he distinguished between the brain and the cerebellum, described the nerves, and identified blood vessels. He also discovered that the pulse originated from the heart, a revolutionary idea at the time.

Erasistratus (310–250 BCE), Herophilos's contemporary, expanded this work by studying the circulatory and nervous systems. He argued that the heart functioned as a pump, pushing blood through veins. While many of his ideas were later proven incorrect, his experimental approach inspired future anatomists.

Roman Advancements: Galen's Influence

The Romans inherited much of their anatomical knowledge from the Greeks. Galen (129–216 CE), a Greek physician working in the Roman Empire, synthesized earlier findings and conducted his own studies. Though his dissections were largely limited to animals (particularly monkeys and pigs), he mapped much of the human body's structure through extrapolation.

Galen distinguished between arteries and veins, demonstrating that arteries carried blood rather than air—a correction to earlier beliefs. However, his understanding of circulation was flawed, as he thought blood was produced in the liver and consumed by the organs. Despite inaccuracies, Galen's writings dominated medical thought in Europe and the Middle East for over a thousand years, forming the basis of medieval and Renaissance medicine.

Medieval Period: Preservation and Expansion of Knowledge

During the Middle Ages, much of the anatomical and biological knowledge from Greek and Roman sources was preserved by Islamic scholars. Figures like Avicenna (980–1037 CE) expanded upon Galenic ideas while introducing innovations of their own. His *Canon of Medicine* synthesized ancient and contemporary knowledge into a comprehensive medical text that remained influential for centuries.

In Europe, the dissection of human cadavers began to reemerge, albeit slowly and under strict supervision by religious authorities. Universities like Bologna became centers of anatomical study, using dissections to verify ancient texts. The practice marked a gradual shift from rote acceptance of ancient authority to empirical observation.

Renaissance: Revolution in Anatomy

The Renaissance brought about a profound transformation in the study of human biology. Artists like Leonardo da Vinci (1452–1519) and Michelangelo (1475–1564) dissected human cadavers to improve their art, inadvertently advancing anatomical knowledge. Leonardo's detailed sketches of muscles, organs, and the vascular system remain some of the most accurate of his time, although they were unpublished during his lifetime.

Andreas Vesalius (1514–1564), a Flemish anatomist, challenged the dominance of Galen's work with his revolutionary book *De humani corporis fabrica* (On the Fabric of the Human Body), published in 1543. Based on meticulous dissections, Vesalius corrected numerous errors in Galenic anatomy, such as the structure of the jawbone and the placement of the heart's ventricles. His insistence on hands-on dissection set a new standard for anatomical studies, marking the beginning of modern human biology.

Scientific Revolution: Circulation and Microscopy

The 17th century brought further breakthroughs in human biology, driven by advances in technology and experimentation. William Harvey (1578–1657) is credited with discovering the full circulation of blood. Through experiments on live animals and careful observation, Harvey demonstrated that the heart acted as a pump, circulating blood through a closed system of arteries and veins. His findings overturned centuries of misconceptions and laid the groundwork for modern cardiovascular physiology.

Around the same time, the invention of the microscope opened new frontiers in biology. Dutch scientist Antonie van Leeuwenhoek (1632–1723) was among the first to observe blood cells, sperm cells, and muscle fibers under a microscope. His discoveries revealed the microscopic structures that underpinned human biology, offering insights into tissues and cells that were previously invisible.

The Enlightenment and Beyond: Physiology and Pathology

During the Enlightenment, biologists began exploring the functions of organs and systems in greater detail. Albrecht von Haller (1708–1777), a Swiss physiologist, conducted pioneering research on muscle and nerve function, establishing the basis for neuromuscular physiology. He demonstrated that nerves carried signals to muscles, advancing understanding of how the body responded to stimuli.

The 19th century saw the emergence of pathology as a distinct field. Rudolf Virchow (1821–1902), often called the "Father of Modern Pathology," emphasized the cellular basis of disease. He proposed that diseases originated from abnormalities in cells, a principle now fundamental to biology and medicine. Virchow's work linked cellular biology with clinical practice, bridging the gap between microscopic findings and patient care.

Modern Era: Genetics, Molecular Biology, and Beyond

The 20th century revolutionized human biology through the discovery of DNA and the advent of molecular biology. In 1953, James Watson and Francis Crick, building on the work of Rosalind Franklin and others, identified the double-helix structure of DNA. This breakthrough explained how genetic information was stored and transmitted, opening the door to genetic engineering and personalized medicine.

Advances in biochemistry and cell biology revealed the intricate processes occurring within cells. Researchers like Albert Claude and Christian de Duve used electron microscopy to explore organelles, uncovering structures like lysosomes and the endoplasmic reticulum. These findings illuminated the cellular machinery responsible for energy production, protein synthesis, and waste removal.

The late 20th century also saw the rise of systems biology, an interdisciplinary approach that examines how different biological components interact within the human body. High-throughput technologies, such as DNA sequencing and proteomics, allowed researchers to study entire genomes and protein networks, providing a holistic view of human biology.

21st Century: The Era of Precision Medicine

Today, human biology continues to evolve with unprecedented speed. The Human Genome Project, completed in 2003, mapped all 3 billion base pairs of human DNA, offering a blueprint for understanding genetic variation and disease susceptibility. This milestone has fueled advancements in gene editing technologies like CRISPR-Cas9, which allow scientists to modify DNA with remarkable precision.

Stem cell research has opened new possibilities for regenerative medicine, enabling the growth of tissues and even organs in the lab. Meanwhile, advances in imaging techniques, such as functional MRI (fMRI), have provided real-time insights into brain activity, shedding light on cognition, emotion, and mental health.

Emerging fields like epigenetics explore how environmental factors influence gene expression, revealing that our DNA is not destiny. By studying these mechanisms, scientists are uncovering links between lifestyle, environment, and long-term health outcomes.

The Role of Human Biology in Medicine and Technology

Human biology serves as the foundation for understanding how the body functions, interacts with its environment, and responds to disease. Medicine and technology rely heavily on this knowledge to develop innovative treatments, diagnostics, and solutions that enhance health and quality of life.

Understanding Disease Mechanisms

Advances in human biology have enabled scientists to uncover the molecular and cellular mechanisms behind diseases. For instance, the identification of DNA mutations in specific genes, such as BRCA1 and BRCA2, has clarified their role in hereditary breast and ovarian cancers. Understanding these pathways allows for the development of targeted therapies, such as PARP inhibitors, which disrupt cancer cell replication.

Similarly, insights into the immune system, including the discovery of T-cells and their subtypes, have revolutionized treatments for autoimmune diseases and cancer. Checkpoint inhibitors, a type of immunotherapy, harness the body's immune response to attack tumors by blocking proteins that suppress immune activity.

Drug Development and Pharmacology

Human biology is critical in drug discovery. Pharmacology relies on understanding the biochemical interactions between drugs and the body. For example, beta-blockers were developed after understanding how adrenaline affects heart rate and blood pressure through beta-adrenergic receptors. By blocking these receptors, beta-blockers reduce cardiac workload, effectively treating hypertension and heart disease.

Advances in molecular biology have led to biologics—drugs derived from living cells. Monoclonal antibodies, such as trastuzumab (Herceptin) for HER2-positive breast cancer, are tailored to target specific proteins in the body, minimizing side effects and improving efficacy. The development of biologics depends on a precise understanding of the human immune system and protein interactions.

Regenerative Medicine

Regenerative medicine has emerged as a transformative field due to breakthroughs in human biology. Stem cell research, for instance, has revealed how pluripotent cells can differentiate into various cell types. These cells are now being used to regenerate damaged tissues, such as cartilage for osteoarthritis or neurons for spinal cord injuries. Researchers are also exploring the use of stem cells to develop lab-grown organs, potentially solving the problem of organ transplant shortages.

Additionally, human biology has enabled bioengineering innovations like 3D bioprinting. Using a combination of biomaterials and cells, scientists have created tissue structures that mimic natural tissues. For example, researchers have printed skin grafts for burn victims and cartilage for joint repair. These breakthroughs depend on a detailed understanding of cellular behavior and tissue architecture.

Diagnostics and Imaging

The development of diagnostic tools is rooted in human biology. Blood tests, for instance, are based on understanding biomarkers—molecules in the blood that indicate health or disease. Elevated levels of C-reactive protein (CRP) signal inflammation, while specific cardiac enzymes, such as troponin, confirm myocardial infarction (heart attack). Advances in molecular biology have introduced genetic testing, allowing for the identification of inherited conditions like cystic fibrosis or Huntington's disease.

Imaging technologies such as MRI, CT scans, and PET scans have become indispensable in medicine. These technologies rely on human biology to interpret the images. For example, fMRI detects changes in blood flow to map brain activity, while CT scans provide detailed views of soft tissues and bones. Understanding tissue density, blood flow, and metabolic processes informs how these tools are used in diagnosis and treatment planning.

Wearable Technology and Monitoring

The intersection of human biology and technology has produced wearable devices that monitor vital signs in real time. Fitness trackers measure heart rate using photoplethysmography (PPG), a technique that detects blood volume changes in the capillaries. Continuous glucose monitors (CGMs) for diabetics use sensors implanted under the skin to measure interstitial glucose levels, enabling precise insulin adjustments.

Wearable technologies are also advancing into therapeutic applications. Smart insoles detect gait abnormalities in Parkinson's patients, while implantable devices like pacemakers and defibrillators regulate cardiac function. These innovations rely on a deep understanding of physiological systems and their feedback mechanisms.

Gene Therapy and CRISPR

Gene therapy represents a monumental leap in medicine, enabled by human biology. By correcting faulty genes, this approach addresses the root cause of genetic disorders. For example, in spinal muscular atrophy, a defective SMN1 gene is replaced with a functional copy using viral vectors. This not only halts disease progression but also restores some lost function.

The CRISPR-Cas9 system has revolutionized gene editing by allowing precise modifications to DNA sequences. This technology has been used to treat sickle cell anemia by editing the gene responsible for abnormal hemoglobin. Human biology provided the blueprint for these advancements, from understanding DNA repair mechanisms to identifying target genes.

Artificial Intelligence and Human Biology

AI applications in medicine rely heavily on human biology. Machine learning algorithms analyze large datasets, such as genetic information or imaging scans, to detect patterns linked to disease. For instance, AI systems trained on retinal scans can identify signs of diabetic retinopathy with remarkable accuracy. Similarly, AI-driven drug discovery platforms predict how molecules will interact with biological targets, accelerating the development of new therapies.

Prosthetics and Biomechanics

Advances in biomechanics have revolutionized prosthetics. Modern prosthetic limbs mimic natural movement by integrating sensors and actuators that respond to nerve signals or muscle contractions. Some devices use myoelectric control, where electrodes detect electrical signals from muscles to move the prosthetic. These innovations are grounded in detailed knowledge of human anatomy and neuromuscular physiology.

Neuroprosthetics, such as cochlear implants for hearing loss, extend these principles to sensory systems. By directly stimulating auditory nerves, cochlear implants restore hearing to individuals with profound deafness. Similarly, retinal implants are being developed to restore vision by stimulating retinal cells or the optic nerve

Tools and Techniques for Studying the Human Body

Studying the human body requires tools and techniques that span molecular, cellular, and systemic levels. Each method contributes unique insights, enabling researchers to unravel the complexities of human biology.

Dissection and Cadaver Studies

Dissection remains one of the oldest and most direct methods for studying human anatomy. Medical students and researchers use cadavers to explore the spatial relationships between organs, muscles, and blood vessels. These studies are critical for understanding anatomical variations, surgical approaches, and injury mechanisms.

In modern anatomy labs, plastination—a process that replaces water and fat in tissues with silicone—preserves specimens for long-term use. This technique provides durable and detailed anatomical models for teaching and research.

Microscopy

Microscopy has been central to human biology since its invention in the 17th century. Light microscopy allows scientists to examine tissues and cells at high magnification. Techniques like histology use specific stains to highlight cellular structures, such as hematoxylin and eosin for nuclei and cytoplasm.

Electron microscopy provides even greater resolution, revealing organelles like mitochondria and ribosomes. Scanning electron microscopy (SEM) captures three-dimensional surface details, while transmission electron microscopy (TEM) reveals internal structures. These tools have been essential for understanding cellular function and pathology.

Molecular Techniques

The study of human biology at the molecular level depends on tools like PCR (polymerase chain reaction), which amplifies DNA sequences for analysis. PCR is widely used in genetic testing, forensics, and infectious disease diagnosis. For instance, real-time PCR detects viral RNA, as seen in COVID-19 testing.

Next-generation sequencing (NGS) has revolutionized genomics by enabling rapid, large-scale DNA sequencing. This technology has been instrumental in identifying disease-associated genes, studying genetic variation, and advancing personalized medicine.

Proteomics, which examines the protein complement of a cell or tissue, uses techniques like mass spectrometry to identify and quantify proteins. These methods reveal how proteins interact and change in response to disease or environmental factors.

Imaging Techniques

Medical imaging tools provide non-invasive ways to study the human body. X-rays, the earliest imaging technology, remain indispensable for visualizing bones and detecting fractures. Advanced modalities like MRI use magnetic fields to generate detailed images of soft tissues, while CT scans create cross-sectional images for diagnosing conditions like tumors or internal bleeding.

Ultrasound, which uses high-frequency sound waves, is particularly valuable for real-time imaging of soft tissues. It is commonly used in obstetrics to monitor fetal development and in cardiology to assess heart function.

Functional Studies

Functional studies examine how systems and organs operate. For example, spirometry measures lung function by assessing airflow during inhalation and exhalation. Electrocardiograms (ECGs) record the heart's electrical activity, providing insights into arrhythmias or ischemic changes.

In neuroscience, techniques like electroencephalography (EEG) and functional MRI (fMRI) track brain activity. EEG records electrical signals from neurons, while fMRI measures changes in blood flow associated with neural activity. These tools are crucial for studying cognition, behavior, and neurological disorders.

Biochemical Assays

Biochemical assays quantify molecules like enzymes, hormones, and metabolites in blood or tissues. For instance, enzyme-linked immunosorbent assays (ELISA) detect specific proteins, such as insulin or antibodies. These tests are widely used in diagnostics, drug development, and research.

Mass spectrometry-based metabolomics analyzes the small molecules involved in metabolism, shedding light on physiological states and disease mechanisms. For example, altered metabolite levels in blood can signal metabolic disorders like diabetes or inborn errors of metabolism.

Organ-on-a-Chip Technology

Organ-on-a-chip systems replicate human tissue and organ functions on a microscale. These devices consist of tiny chambers lined with human cells that mimic physiological processes, such as blood flow or lung expansion. Researchers use them to test drug effects, study disease mechanisms, and reduce reliance on animal models.

CRISPR and Gene Editing Tools

CRISPR-Cas9 has become an indispensable tool for studying genes and their functions. By precisely editing specific DNA sequences, scientists can investigate how mutations affect proteins, cells, or entire biological systems. For example, researchers use CRISPR to create cellular models of diseases such as cystic fibrosis by introducing specific mutations. These models help scientists explore disease mechanisms and test potential treatments in a controlled environment.

CRISPR also enables functional genomics studies, where individual genes are "knocked out" to determine their roles in biological processes. This approach has been crucial for identifying drug targets, understanding gene regulation, and exploring the genetic basis of human biology.

Flow Cytometry

Flow cytometry is a powerful technique for analyzing individual cells in a fluid suspension. This tool measures physical and chemical characteristics, such as size, granularity, and the presence of specific proteins on the cell surface. By tagging proteins with fluorescent markers, researchers can identify and isolate cell populations, such as T-cells, stem cells, or cancer cells.

Applications include monitoring immune responses, diagnosing hematological diseases, and studying cellular heterogeneity. Flow cytometry is particularly useful in immunology and oncology, where understanding cell behavior is essential for developing treatments.

Single-Cell Analysis

Single-cell technologies have advanced our understanding of cellular diversity. Unlike bulk analysis, which averages data across many cells, single-cell techniques analyze individual cells, revealing differences in gene expression, protein levels, or metabolic states. Methods like single-cell RNA sequencing (scRNA-seq) allow researchers to map cell types and states in tissues, uncovering previously unrecognized cellular subpopulations.

For example, single-cell analysis has been used to identify rare immune cells involved in cancer immunotherapy or to track how cells change during embryonic development. This approach is transforming fields like developmental biology, cancer research, and regenerative medicine.

Tissue Engineering Techniques

Tissue engineering integrates human biology and engineering principles to create functional tissues. Techniques such as scaffold fabrication involve designing materials that mimic the extracellular matrix, providing structural support for cell growth. Hydrogels, polymers, and bioceramics are commonly used as scaffolds to guide tissue regeneration.

Bioreactors simulate the physiological environment, providing nutrients, oxygen, and mechanical stimuli to engineered tissues. For instance, researchers use bioreactors to grow cartilage, heart valves, or skin equivalents for transplantation or drug testing. These methods rely on understanding cellular behavior, tissue mechanics, and material science.

High-Throughput Screening

High-throughput screening (HTS) involves testing thousands of compounds or genetic modifications in parallel to identify biological effects. Robotic systems and automated assays allow researchers to rapidly evaluate how different molecules interact with proteins, cells, or tissues. HTS is widely used in drug discovery to identify potential therapeutic candidates.

Advances in human biology have made HTS more sophisticated, incorporating complex cellular models or 3D tissue cultures. This improves the relevance of screening results to human physiology, reducing the likelihood of failure in later stages of drug development.

Omics Technologies

Omics technologies—genomics, transcriptomics, proteomics, and metabolomics—provide comprehensive data on the molecular components of human biology. Genomics focuses on DNA sequences, while transcriptomics examines RNA expression levels. Proteomics and metabolomics explore protein abundance and metabolic pathways, respectively.

Integrating these datasets enables systems biology approaches, where researchers study how molecular networks interact to maintain health or contribute to disease. For example, multi-omics studies have revealed how genetic mutations, protein dysfunction, and metabolic imbalances converge in diseases like diabetes or Alzheimer's.

Advances in Imaging at the Cellular Level

Recent innovations have pushed imaging technology beyond traditional light and electron microscopy. Super-resolution microscopy, such as STED (stimulated emission depletion) and PALM (photoactivated localization microscopy), breaks the diffraction limit of light, allowing researchers to visualize structures at the nanometer scale.

Live-cell imaging enables scientists to observe cellular processes in real time, providing insights into dynamic events like cell division, migration, or intracellular transport. Fluorescent reporters, such as GFP (green fluorescent protein), label specific molecules, making them visible under microscopes.

These advanced imaging techniques reveal the intricate architecture and behavior of cells, bridging the gap between structure and function in human biology.

Computational Biology and Modeling

Computational tools are indispensable for studying human biology in the digital age. Simulations and models predict how biological systems behave under different conditions. For example, researchers use computational models to simulate blood flow in arteries, predict drug interactions, or study the spread of infectious diseases.

Machine learning algorithms analyze complex datasets, identifying patterns that might be imperceptible to humans. For example, AI models trained on gene expression data can predict how certain genes influence disease outcomes. Computational biology complements experimental methods, accelerating discoveries in human biology.

Integrative Approaches to Human Biology

Combining multiple tools and techniques provides a holistic understanding of human biology. For instance, studying cancer requires integrating molecular data (e.g., genetic mutations), cellular information (e.g., tumor microenvironment), and systemic insights (e.g., immune responses). By leveraging diverse methodologies, researchers gain a more complete picture of health and disease.

These integrative approaches are especially valuable in emerging fields like personalized medicine and systems biology, where the complexity of the human body demands cross-disciplinary collaboration and innovation.

The Importance of Homeostasis in Maintaining Life

Homeostasis is the body's ability to maintain a stable internal environment despite changes in external conditions. This regulation is vital for survival because cells and biochemical processes require specific conditions to function optimally. Disruptions in homeostasis can lead to disease or death, highlighting its importance in human biology.

Temperature Regulation

One of the most critical aspects of homeostasis is thermoregulation. The human body must maintain a core temperature around 37°C (98.6°F) for enzymes and cellular processes to function properly. When exposed to heat, the hypothalamus in the brain activates mechanisms such as sweating and vasodilation. Sweat glands release moisture onto the skin, which cools the body as it evaporates. Vasodilation increases blood flow to the skin, allowing excess heat to dissipate.

In cold conditions, the body conserves heat through vasoconstriction, which reduces blood flow to the skin, and shivering, which generates heat by rapid muscle contractions. Brown adipose tissue, found primarily in infants and some adults, also produces heat through non-shivering thermogenesis by metabolizing fat.

Failure in thermoregulation, such as during heatstroke or hypothermia, can disrupt enzyme function and damage vital organs.

Blood Glucose Regulation

Maintaining stable blood glucose levels is essential for energy supply to cells, especially in the brain. The endocrine system is key in this process. When blood glucose rises after a meal, the pancreas secretes insulin, prompting cells to take in glucose for energy or storage as glycogen in the liver and muscles. If glucose levels drop, the pancreas releases glucagon, which signals the liver to break down glycogen into glucose and release it into the bloodstream.

Chronic disruption of glucose homeostasis can result in diabetes. In type 1 diabetes, the body cannot produce insulin, while in type 2 diabetes, cells become resistant to insulin's effects. Both conditions highlight the importance of tightly regulated homeostasis for long-term health.

Fluid and Electrolyte Balance

The body's water and electrolyte levels are crucial for maintaining cell function and blood pressure. Osmoregulation ensures that cells neither shrink nor swell excessively due to imbalances in water concentration. The kidneys regulate water levels by adjusting urine concentration based on signals from hormones like antidiuretic hormone (ADH). When dehydrated, ADH increases water reabsorption in the kidneys, reducing urine output.

Electrolytes, such as sodium, potassium, and calcium, are vital for nerve conduction, muscle contraction, and heart function. The body's sodium-potassium pump maintains proper ion gradients across cell membranes, which are essential for generating electrical signals. Disorders like hyponatremia (low sodium) or hyperkalemia (high potassium) can disrupt these processes, leading to life-threatening complications.

Acid-Base Balance

Human cells operate within a narrow pH range of approximately 7.35 to 7.45. Maintaining this balance is essential for biochemical reactions and protein stability. The body uses buffer systems, such as the bicarbonate buffer, to resist pH changes. The respiratory system regulates pH by adjusting carbon dioxide levels, while the kidneys excrete hydrogen ions and reabsorb bicarbonate to maintain balance.

Acidosis, an excessive drop in pH, can impair enzyme activity and reduce oxygen delivery to tissues. Conversely, alkalosis, an increase in pH, can disrupt ion balance and nerve function. Both conditions demonstrate the critical nature of pH homeostasis.

Oxygen and Carbon Dioxide Regulation

The respiratory system ensures that oxygen levels remain sufficient for cellular respiration while eliminating carbon dioxide, a byproduct of metabolism. Chemoreceptors in the brainstem monitor blood oxygen and carbon dioxide levels. When carbon dioxide levels rise, the body increases breathing rate and depth to expel it. Oxygen homeostasis is maintained through hemoglobin, which transports oxygen in red blood cells.

Disruptions in this system, such as during hypoxia (low oxygen levels) or hypercapnia (high carbon dioxide levels), can quickly impair cellular metabolism and lead to organ failure.

Hormonal Regulation

Endocrine feedback loops are central to maintaining homeostasis. For example, the hypothalamus and pituitary gland regulate thyroid hormone levels through the hypothalamic-pituitary-thyroid axis. This ensures appropriate metabolic rates for energy production and temperature regulation. Similarly, the hypothalamic-pituitary-adrenal axis governs cortisol levels, which help the body respond to stress and regulate metabolism.

Dysregulation of these systems, such as in hyperthyroidism or Addison's disease, underscores the importance of precise hormonal control for homeostasis.

Human Biology in Evolutionary Context: From Primates to Homo sapiens

The evolutionary history of humans provides critical insights into human biology. Understanding how modern Homo sapiens evolved from primate ancestors highlights the adaptations that shaped our anatomy, physiology, and behavior.

Primate Origins and Adaptations

Humans belong to the order Primates, a group that includes monkeys, apes, and humans. Primates evolved approximately 65 million years ago, adapting to arboreal (tree-dwelling) environments. Key adaptations included forward-facing eyes for depth perception, grasping hands with opposable thumbs for climbing and manipulating objects, and large brains relative to body size for problem-solving.

Dietary versatility also emerged during this period. Early primates were primarily frugivorous, but many developed the ability to consume leaves, insects, and other foods. This dietary flexibility laid the groundwork for later dietary shifts in human evolution.

Bipedalism: A Defining Trait

Around 7 million years ago, the first hominins—human ancestors—diverged from the lineage leading to modern chimpanzees. One of the earliest defining traits of hominins was bipedalism. Fossil evidence from species like *Australopithecus afarensis* (e.g., the famous "Lucy" specimen) shows adaptations for upright walking, including a more vertical spine, a wider pelvis, and arched feet.

Bipedalism freed the hands for tool use and carrying objects, providing significant evolutionary advantages. It also reduced energy expenditure compared to quadrupedal locomotion, enabling hominins to travel long distances efficiently.

Tool Use and Cognitive Expansion

The genus Homo, which includes modern humans, emerged around 2.5 million years ago with species like *Homo habilis*. This period marked the first clear evidence of tool use. Early stone tools, such as Oldowan flakes, were used for cutting and processing food. Tool use drove changes in hand anatomy, including a more robust thumb and refined grip, which further facilitated technological innovation.

Cognitive expansion also characterized the Homo lineage. Brain size increased significantly, from about 450 cubic centimeters in *Australopithecus* to over 1,400 cubic centimeters in modern humans. This growth enabled more complex social structures, communication, and problem-solving abilities.

Dietary Shifts and Evolutionary Impacts

Diet has been critical in human evolution. The transition from primarily plant-based diets to omnivory, including meat consumption, provided high-quality nutrients that supported brain growth. The advent of cooking, likely practiced by *Homo erectus*, made food easier to digest, increasing energy availability. This shift reduced the need for large digestive tracts, freeing more energy for brain development.

Dietary adaptations are evident in human biology. For example, the evolution of smaller jaws and teeth reflects a transition to softer, processed foods. Additionally, the persistence of lactase enzyme activity in some human populations, enabling the digestion of dairy products into adulthood, illustrates an adaptation to agricultural lifestyles.

Speech and Language

The capacity for complex speech is a hallmark of Homo sapiens. Anatomical changes in the larynx and vocal tract allowed for a wider range of sounds, while neural developments in regions like Broca's and Wernicke's areas enabled language processing. The FOXP2 gene, associated with speech and language, underwent specific mutations in the human lineage, further facilitating these abilities.

Language provided significant evolutionary advantages, enabling cooperation, cultural transmission, and the development of sophisticated tools and social structures.

Neanderthals and Interbreeding

Modern humans share a close relationship with Neanderthals (*Homo neanderthalensis*), who lived in Europe and Asia until about 40,000 years ago. Genetic evidence shows that interbreeding occurred between Neanderthals and early Homo sapiens. Approximately 1–2% of the genome in modern non-African populations comes from Neanderthals.

These genetic contributions influence traits such as immune system function and skin pigmentation, illustrating how interbreeding shaped human biology.

Adaptations to Environment

As Homo sapiens spread out of Africa around 70,000 years ago, they adapted to diverse environments. Populations in colder climates developed traits like shorter limbs and increased body fat for heat retention (Allen's and Bergmann's rules), while those in tropical regions retained leaner builds for efficient cooling.

Skin pigmentation evolved as an adaptation to ultraviolet (UV) radiation. Darker skin protects against UV-induced DNA damage and folate degradation in high-UV environments, while lighter skin enhances vitamin D synthesis in low-UV regions. These variations underscore the interplay between biology and environmental pressures.

Modern Evolutionary Changes

Human evolution is ongoing. The spread of genetic traits such as resistance to infectious diseases (e.g., malaria resistance via sickle cell trait) demonstrates recent adaptations to environmental challenges. Additionally, cultural and technological advances continue to shape selective pressures, altering the trajectory of human evolution.

The Human Microbiome: The Body's Hidden Ecosystem

The human microbiome encompasses the vast collection of microorganisms—bacteria, viruses, fungi, and archaea—that live on and within the human body. These microscopic inhabitants outnumber human cells by approximately 10 to 1 and are integral to health, influencing processes ranging from digestion to immunity.

Composition and Diversity of the Microbiome

The human microbiome is most densely populated in specific regions, including the gut, skin, oral cavity, respiratory tract, and urogenital system. The gut microbiome, concentrated in the colon, is the most diverse and studied. Over 1,000 bacterial species inhabit the gut, with dominant groups including *Firmicutes*, *Bacteroidetes*, *Actinobacteria*, and *Proteobacteria*. Each individual's microbiome is unique, influenced by genetics, diet, environment, and lifestyle.

In the skin microbiome, microorganisms like *Staphylococcus epidermidis* and *Propionibacterium acnes* help protect against pathogenic bacteria. In the mouth, *Streptococcus* species prevent harmful bacteria from taking hold, while in the vaginal microbiome, *Lactobacillus* species maintain an acidic environment that limits infection.

Roles in Digestion and Metabolism

The gut microbiome is important in breaking down complex carbohydrates, fibers, and other indigestible compounds. Bacteria like *Bacteroides* and *Ruminococcus* ferment these substances, producing short-chain fatty acids (SCFAs) such as acetate, butyrate, and propionate. SCFAs are absorbed by intestinal cells and used for energy, while also contributing to gut health by reducing inflammation and strengthening the intestinal barrier.

The microbiome also synthesizes essential vitamins, including vitamin K and certain B vitamins, such as B12 and folate. These contributions are critical, as the human body cannot produce these nutrients independently.

Immune System Interactions

The microbiome educates and regulates the immune system, helping the body distinguish between harmful pathogens and harmless microorganisms. Gut bacteria produce metabolites that influence the activity of immune cells like T-regulatory cells, which maintain immune tolerance and prevent overactive responses.

During early development, exposure to a diverse microbiome is essential for building a robust immune system. A lack of microbial exposure has been linked to an increased risk of autoimmune diseases, such as type 1 diabetes and Crohn's disease, supporting the "hygiene hypothesis." This theory suggests that modern sanitation and reduced microbial exposure contribute to immune system dysregulation.

Influence on the Brain and Behavior

The gut-brain axis is a bidirectional communication network connecting the gut microbiome and the central nervous system. Gut bacteria produce neurotransmitters like serotonin, dopamine, and gamma-aminobutyric acid (GABA), which influence mood and cognition. For instance, about 90% of the body's serotonin is produced in the gut.

Emerging evidence links microbiome imbalances, or dysbiosis, to mental health conditions like depression and anxiety. Studies have shown that restoring microbial balance through probiotics or dietary changes can alleviate symptoms in some cases.

Microbiome and Disease

Dysbiosis is implicated in a wide range of diseases. Inflammatory bowel diseases (IBD), such as Crohn's disease and ulcerative colitis, are associated with reduced microbial diversity and an overgrowth of harmful bacteria. Similarly, metabolic conditions like obesity and type 2 diabetes have been linked to an altered gut microbiome that promotes energy extraction and fat storage.

The microbiome also influences cancer risk. Certain gut bacteria metabolize dietary components into carcinogenic compounds, while others produce anti-inflammatory metabolites that protect against cancer. For example, *Fusobacterium nucleatum* is associated with colorectal cancer, whereas *Lactobacillus* species may have protective effects.

Personalized Medicine and Microbiome Research

The microbiome is at the forefront of personalized medicine. Fecal microbiota transplants (FMT) are already used to treat recurrent *Clostridioides difficile* infections, with success rates exceeding 90%. Researchers are also developing microbiome-based therapies for autoimmune diseases, allergies, and even neurodegenerative conditions.

Advances in sequencing technology have enabled scientists to map the microbiome in unprecedented detail. These insights are guiding interventions such as tailored probiotic formulations, dietary recommendations, and microbiome-targeted drugs.

Ethics in Human Biology: Challenges and Considerations

Ethical considerations in human biology are integral to ensuring that advancements in the field benefit humanity while minimizing harm. As technologies and research methods evolve, so do the ethical dilemmas they pose, necessitating careful reflection and regulation.

Informed Consent and Autonomy

In human biology research, obtaining informed consent is a cornerstone of ethical practice. Participants must fully understand the purpose, risks, and benefits of a study before agreeing to take part. However, advances in genetic testing and biobanking complicate this process. For example, participants may consent to donate their genetic material for a specific study, but the data could later be used for unrelated purposes.

Autonomy is also challenged by technologies like CRISPR-Cas9, which can edit the genomes of embryos. While this has potential to eliminate genetic disorders, it raises questions about consent for individuals whose genetic makeup is altered before birth. Ensuring that interventions respect individual rights is a complex ethical challenge.

Privacy and Genetic Data

The collection and storage of genetic data present significant privacy concerns. Genetic information can reveal sensitive details about an individual's health, ancestry, and predisposition to diseases. If mishandled, this data could lead to discrimination in employment, insurance, or other areas.

The rise of direct-to-consumer genetic testing services, such as 23andMe, exacerbates these risks. Consumers often share their genetic data without fully understanding how it might be used or shared with third parties. Regulatory frameworks are essential to safeguard genetic privacy and prevent misuse.

Equity and Access

Advancements in human biology, particularly in precision medicine, often come with high costs, making them inaccessible to many. For example, gene therapies for rare diseases can cost millions of dollars, limiting their availability to those in wealthier countries or with substantial financial resources. This creates ethical dilemmas around health equity and the distribution of medical advancements.

The disparity extends to research as well. Clinical trials often underrepresent marginalized populations, leading to treatments that may not be effective or safe for all demographic groups. Addressing these disparities requires intentional efforts to include diverse populations in research and ensure equitable access to innovations.

Dual-Use Dilemmas

Some advancements in human biology have dual-use potential, meaning they can be used for both beneficial and harmful purposes. For example, synthetic biology can create lifesaving treatments, but the same techniques could be used to engineer bioweapons. Balancing innovation with security is a persistent ethical challenge in fields like genetic engineering and microbiome research.

Similarly, neurobiology research into brain-computer interfaces has applications in restoring mobility for paralyzed individuals but could also be misused for surveillance or behavioral control. Ethical frameworks must anticipate and address such dual-use risks.

Animal and Human Research Ethics

Research in human biology often involves animal models to study diseases and test treatments. While these studies provide insights, they raise ethical concerns about animal welfare. Alternatives like organ-on-a-chip technology and computer modeling are being developed to reduce reliance on animal testing, but these methods are not yet comprehensive.

Human experimentation poses additional challenges. Historical abuses, such as the Tuskegee Syphilis Study, underscore the importance of strict ethical oversight. Institutional review boards (IRBs) and international guidelines, like the Declaration of Helsinki, aim to prevent exploitation and ensure that human research is conducted responsibly.

Genome Editing and Germline Modifications

The ability to edit human genomes has sparked significant ethical debates. Germline editing, which alters DNA in eggs, sperm, or embryos, has the potential to eliminate genetic diseases but also introduces risks of unintended consequences. Off-target effects could lead to new health issues, and edited traits would be passed on to future generations, raising concerns about long-term impacts.

Beyond medical uses, germline editing could enable genetic enhancements, such as increased intelligence or physical ability. This raises questions about societal implications, including exacerbating inequality and altering perceptions of normalcy. Establishing clear boundaries for the use of genome editing is critical to addressing these ethical challenges.

Ownership of Biological Data and Materials

As human biology research increasingly relies on biobanks and genomic databases, questions about ownership arise. Who owns the biological samples or data once they are collected? Participants often have little control over how their contributions are used, even though they may contain highly personal information.

This issue extends to commercial applications. For example, pharmaceutical companies frequently use public genetic databases to develop profitable treatments, raising concerns about whether individuals or communities should receive compensation for their contributions.

Cultural and Religious Considerations

Ethical perspectives on human biology are not universal. Cultural and religious beliefs often influence views on practices like stem cell research, organ transplantation, and genetic modification. For example, some cultures oppose the use of embryonic stem cells because it involves the destruction of embryos, while others may object to certain interventions based on spiritual or traditional principles.

Respecting these diverse perspectives while advancing scientific research requires careful dialogue and compromise. Researchers must engage with communities to ensure that their work aligns with broader societal values.

Ethical Oversight and Public Engagement

The rapid pace of innovation in human biology often outstrips the development of ethical guidelines. Establishing effective oversight mechanisms is essential to ensure that research and applications align with societal values. Public engagement is a key component of this process, as it allows diverse voices to contribute to the ethical discourse.

Transparent communication about risks, benefits, and uncertainties helps build trust between scientists and the public. This is especially important in controversial areas like gene editing and microbiome manipulation, where societal acceptance is critical for progress.

CHAPTER 2: THE BUILDING BLOCKS OF LIFE

The Chemistry of Life: Atoms, Molecules, and Biomolecules

Life is built from a rich array of chemical elements and compounds that interact in precise ways to sustain biological functions. At its core, the chemistry of life hinges on the properties of atoms and the molecules they form. Understanding these components offers a foundation for grasping how cells, tissues, and organs work.

Atoms: The Basic Units of Matter

Atoms are the smallest units of matter that retain the properties of an element. They consist of three primary subatomic particles: protons, neutrons, and electrons. Protons, which are positively charged, and neutrons, which carry no charge, form the dense nucleus at the center of the atom. Electrons, negatively charged particles, orbit the nucleus in defined energy levels or shells.

The number of protons in the nucleus defines an element's atomic number. For example, hydrogen, the simplest element, has one proton, while carbon has six. Electrons, arranged in shells around the nucleus, determine how an atom interacts with other atoms. Atoms are most stable when their outermost shell is full, and they tend to gain, lose, or share electrons to achieve this state.

Elements Essential to Life

Of the 118 known elements, only a small subset is necessary for life. **Carbon, hydrogen, oxygen, and nitrogen** are the most abundant, making up about 96% of the human body. These elements are versatile and reactive, forming the backbone of biomolecules.

- Carbon is unparalleled in its ability to form long chains and complex structures, thanks to its four valence electrons. This makes it the key element in organic molecules.
- Oxygen is critical for cellular respiration and is a major component of water, which constitutes about 60% of the human body.
- Hydrogen, the lightest and most abundant element, is essential in forming water and organic molecules.
- Nitrogen is a key component of amino acids and nucleotides, the building blocks of proteins and DNA.

Other essential elements include phosphorus, which is vital for DNA and ATP (the energy currency of the cell), and sulfur, which stabilizes protein structures through disulfide bonds. Trace elements like iron, zinc, and iodine are required in small amounts but are crucial for specific biological functions, such as oxygen transport and enzyme activity.

Chemical Bonds and Molecule Formation

Atoms interact through chemical bonds to form molecules, which are the building blocks of matter. There are three primary types of bonds: ionic, covalent, and hydrogen bonds.

- **Ionic bonds** occur when one atom transfers electrons to another, creating charged particles called ions. For example, sodium (Na) donates an electron to chlorine (Cl), forming sodium chloride (NaCl), or table salt.
- **Covalent bonds** involve the sharing of electrons between atoms, forming strong and stable molecules. Water (H_2O), where two hydrogen atoms share electrons with one oxygen atom, is an example of a covalent molecule.
- **Hydrogen bonds** are weaker but highly significant in biology. These bonds form between polar molecules, such as between water molecules or within DNA strands, where they stabilize the helical structure.

These bonds determine the shape and reactivity of molecules, which directly influence their biological functions.

Water: The Solvent of Life

Water is essential for life, acting as a solvent, a medium for chemical reactions, and a temperature buffer. Its unique properties stem from its polar nature. The oxygen atom in a water molecule pulls electrons more strongly than the hydrogen atoms, creating partial charges. This polarity allows water to dissolve ionic compounds like salts and interact with other polar molecules, making it a universal solvent.

Water's ability to form hydrogen bonds gives it high cohesion and surface tension, allowing it to move through narrow spaces, such as capillaries in plants and blood vessels in humans. Its high specific heat helps regulate temperature, protecting organisms from rapid environmental changes.

Biomolecules: The Molecules of Life

Biomolecules are large, complex molecules that perform essential biological functions. They are categorized into four main types: carbohydrates, lipids, proteins, and nucleic acids.

1. Carbohydrates: Quick Energy and Structural Support

Carbohydrates consist of carbon, hydrogen, and oxygen in a 1:2:1 ratio. They are the primary source of energy for the body. The simplest carbohydrates, monosaccharides, include glucose, fructose, and galactose. Glucose is the body's main energy source, broken down during cellular respiration to produce ATP.

Monosaccharides combine to form disaccharides (e.g., sucrose, or table sugar) and polysaccharides (e.g., starch, glycogen, and cellulose). Glycogen, stored in the liver and muscles, provides a quick energy reserve for the body. Cellulose, found in plant cell walls, is indigestible by humans but serves as dietary fiber, aiding digestion.

2. Lipids: Energy Storage and Membrane Structure

Lipids are hydrophobic molecules composed mainly of carbon and hydrogen. They include fats, oils, phospholipids, and steroids. Lipids serve multiple functions: long-term energy storage, insulation, and forming cell membranes.

- **Triglycerides**, made of glycerol and three fatty acids, are the main form of fat storage. Saturated fats have no double bonds in their fatty acid chains, while unsaturated fats have one or more double bonds, creating kinks that affect their properties.
- **Phospholipids** are essential for cell membranes. They have a hydrophilic head and hydrophobic tails, forming a bilayer that regulates what enters and exits cells.
- **Steroids**, like cholesterol, are precursors to hormones such as estrogen and testosterone.

Lipids are calorie-dense, providing about 9 calories per gram, compared to 4 calories per gram for carbohydrates and proteins.

3. Proteins: Workhorses of the Cell

Proteins are polymers of amino acids, linked by peptide bonds. The 20 standard amino acids vary in their side chains, giving proteins diverse structures and functions. Protein structure has four levels of organization:

- **Primary structure:** The linear sequence of amino acids.
- **Secondary structure:** Folding into alpha-helices or beta-sheets, stabilized by hydrogen bonds.
- **Tertiary structure:** The overall 3D shape of a single polypeptide.
- **Quaternary structure:** The assembly of multiple polypeptides, as seen in hemoglobin.

Proteins perform countless roles, including enzyme catalysis (e.g., amylase breaks down starch), transport (e.g., hemoglobin carries oxygen), signaling (e.g., insulin regulates blood glucose), and structural support (e.g., collagen in connective tissues).

4. Nucleic Acids: Information Storage and Transfer

Nucleic acids, DNA and RNA, store and transmit genetic information. They are composed of nucleotides, each consisting of a sugar, a phosphate group, and a nitrogenous base.

- **DNA (deoxyribonucleic acid)** contains the instructions for building proteins. Its double-helix structure, stabilized by hydrogen bonds between complementary bases (adenine-thymine, cytosine-guanine), ensures the fidelity of genetic information.
- **RNA (ribonucleic acid)** acts as a messenger, translating DNA instructions into proteins. RNA differs from DNA by having ribose sugar and uracil instead of thymine.

The sequence of nucleotides in DNA forms genes, which are expressed through transcription (DNA to RNA) and translation (RNA to protein).

Enzymes: Catalysts of Life

Enzymes, a subset of proteins, speed up biochemical reactions by lowering the activation energy required. Each enzyme is specific to a substrate, binding at an active site like a lock and key. For example, lactase breaks down lactose into glucose and galactose. Without enzymes, most reactions in the body would proceed too slowly to sustain life.

Chemical Energy: ATP

Adenosine triphosphate (ATP) is the energy currency of cells. It consists of adenine, ribose, and three phosphate groups. The energy stored in the bonds between phosphate groups is released during hydrolysis, fueling cellular processes like muscle contraction, nerve impulses, and molecule transport.

Overview of Organic Molecules: Proteins, Carbohydrates, Lipids, and Nucleic Acids

Organic molecules form the structural and functional framework of life. These compounds are primarily composed of carbon, hydrogen, and oxygen, with other elements like nitrogen, phosphorus, and sulfur playing critical roles. The four major classes of organic molecules—proteins, carbohydrates, lipids, and nucleic acids—are indispensable for cellular structure, metabolism, and genetic inheritance.

Proteins are the most versatile biomolecules. They are composed of amino acids linked together by peptide bonds, forming chains that fold into complex three-dimensional shapes. These structures allow proteins to serve numerous functions. Enzymes, a type of protein, catalyze biochemical reactions, speeding up processes that would otherwise occur too slowly to sustain life. For example, DNA polymerase helps replicate DNA during cell division. Structural proteins like keratin in hair and nails or collagen in connective tissues provide mechanical support, while transport proteins, such as hemoglobin, carry oxygen throughout the body. Signal proteins like insulin regulate physiological processes by mediating communication between cells. Proteins are synthesized based on the genetic instructions encoded in DNA, ensuring their structures are finely tuned for specific tasks.

Carbohydrates provide energy and structural components. They consist of monosaccharides, disaccharides, and polysaccharides. Glucose, a monosaccharide, is central to cellular respiration, where it is oxidized to generate ATP, the energy currency of cells. Disaccharides like lactose and sucrose are formed by linking two monosaccharides, and they serve as short-term energy sources. Polysaccharides, including starch, glycogen, and cellulose, are long chains of glucose molecules. Glycogen acts as an energy reserve stored in the liver and muscles, while cellulose forms the rigid cell walls of plants. Although humans cannot digest cellulose, it aids digestion as dietary fiber. Carbohydrates also contribute to cell recognition and communication, with glycoproteins on cell surfaces playing key roles in immune responses and cell signaling.

Lipids are hydrophobic molecules that include fats, phospholipids, and steroids. Triglycerides, the most common fats, consist of glycerol bonded to three fatty acid chains. They serve as long-term energy storage, providing more than twice the energy per gram as carbohydrates or proteins. Fat also insulates the body and cushions vital organs. Phospholipids, which contain a hydrophilic head and two hydrophobic tails, are essential for cell membranes. They arrange themselves into bilayers, forming a barrier that controls the movement of substances into and out of cells. Steroids, such as cholesterol, are another type of lipid. Cholesterol is a precursor to hormones like estrogen and testosterone and is also a component of cell membranes, contributing to their fluidity and stability.

Nucleic acids, DNA and RNA, **store and transmit genetic information**. DNA, a double-helix structure composed of nucleotides, encodes instructions for protein synthesis. Each nucleotide includes a sugar, a phosphate group, and one of four nitrogenous bases: adenine, thymine, cytosine, or guanine. Complementary base pairing (A-T, C-G) ensures accurate replication during cell division. RNA, single-stranded and containing uracil instead of thymine, translates DNA instructions into functional proteins. Messenger RNA (mRNA) carries genetic codes from the nucleus to ribosomes, where proteins are assembled, while transfer RNA (tRNA) and ribosomal RNA (rRNA) facilitate the process. Nucleic acids also play roles beyond genetics; for instance, ATP, a modified nucleotide, powers cellular activities.

The interaction of these organic molecules creates the biochemical pathways necessary for growth, reproduction, and survival. Their unique properties and interdependence illustrate the complexity of life at the molecular level.

The Role of Water in Human Biology

Water is fundamental to human biology, constituting about 60% of an adult's body weight. It is involved in nearly every biological process, from molecular interactions to organ function. Water's unique properties, including its polarity, solvent capabilities, and thermal stability, make it indispensable for maintaining life.

As a polar molecule, water facilitates countless biochemical reactions. The oxygen atom in a water molecule pulls electrons more strongly than the hydrogen atoms, creating partial charges. This polarity allows water to dissolve ionic and polar compounds, forming aqueous solutions where reactions occur. Enzymes, substrates, and ions interact effectively in this medium, driving processes like metabolism, DNA replication, and energy production. Nonpolar molecules, such as lipids, do not dissolve in water, which leads to the formation of cellular membranes. The hydrophobic interactions of lipid molecules are a direct result of water's chemical behavior, creating compartments essential for life.

Water is critical for **transporting substances within the body**. Blood plasma, which is 90% water, carries nutrients, hormones, and oxygen to cells while removing waste products like carbon dioxide and urea. Water's high heat capacity stabilizes the temperature of blood, ensuring that biochemical reactions occur within a narrow thermal range. Lymphatic fluid, another water-based medium, transports immune cells and absorbs dietary fats in the digestive system. In addition, cerebrospinal fluid cushions the brain and spinal cord, providing both physical protection and a means for nutrient exchange.

Thermoregulation depends heavily on water. Sweating and evaporative cooling are the body's primary responses to heat stress. As sweat glands release water onto the skin, the evaporation process removes heat, lowering body temperature. This mechanism prevents overheating, which could denature proteins and disrupt cellular function. Similarly, water's high specific heat capacity helps buffer temperature fluctuations, maintaining a stable internal environment despite external changes.

In the digestive system, water acts as both a solvent and a lubricant. It dissolves nutrients, allowing enzymes to break them down into absorbable units. Saliva, which is mostly water, contains enzymes like amylase that begin carbohydrate digestion. Water also lubricates food for easier swallowing and facilitates the movement of chyme through the gastrointestinal tract. In the intestines, water aids in nutrient absorption, transporting molecules across the intestinal lining into the bloodstream.

Water maintains cellular structure and function. Inside cells, the cytoplasm is a water-based solution where organelles are suspended and metabolic reactions occur. The presence of water ensures that macromolecules like proteins and nucleic acids maintain their proper shapes and functions. For example, hydrogen bonds between water molecules stabilize the helical structure of DNA and the folding of proteins. Water also generates osmotic pressure, regulating the movement of substances across cell membranes. This prevents cells from shrinking or swelling excessively, preserving their integrity.

Water contributes to waste elimination. The kidneys filter blood to remove metabolic byproducts, producing urine, which is mostly water. This process relies on water to dilute toxins, ensuring they are safely excreted. Dehydration reduces kidney efficiency, leading to concentrated urine and an increased risk of kidney stones or urinary tract infections. Water also assists in expelling waste through sweat and respiration, where it carries heat and gases like carbon dioxide out of the body.

Acid-base balance, critical for enzyme activity and cellular function, is tightly regulated by water. The bicarbonate buffer system, which maintains blood pH around 7.4, depends on water to facilitate chemical equilibria. Carbon dioxide reacts with water to form carbonic acid, which dissociates into hydrogen ions and bicarbonate. This reversible reaction helps neutralize excess acids or bases, preventing harmful shifts in pH.

Water's role **extends to reproduction and development**. Amniotic fluid, primarily water, surrounds and cushions the developing fetus, providing a stable environment for growth. It also facilitates nutrient and waste exchange between the mother and fetus. During childbirth, water in the amniotic sac aids in lubrication, reducing friction and supporting delivery.

The unique properties of water—its polarity, cohesion, and ability to form hydrogen bonds—enable it to support the complex processes of human biology. Without water, the intricate balance of biochemical reactions, cellular activities, and systemic functions that sustain life would not be possible.

CHAPTER 3: DNA AND GENETICS

Structure and Function of DNA: The Blueprint of Life

DNA, or deoxyribonucleic acid, is the molecule that stores genetic information in all living organisms. It directs the development, functioning, growth, and reproduction of cells. Every cell in the human body, except for red blood cells, contains DNA. Its structure and function are finely tuned to manage the vast amount of information needed to sustain life.

The Double Helix: Structure of DNA

DNA has a double-helix structure, first described by James Watson and Francis Crick in 1953, based on Rosalind Franklin's X-ray diffraction data. The double helix resembles a twisted ladder, with two strands running in opposite directions. These strands are composed of nucleotides, the basic building blocks of DNA.

Each nucleotide consists of three components: a phosphate group, a sugar molecule (deoxyribose), and one of four nitrogenous bases. The nitrogenous bases—adenine (A), thymine (T), cytosine (C), and guanine (G)—are paired specifically through hydrogen bonding. Adenine pairs with thymine via two hydrogen bonds, while cytosine pairs with guanine via three hydrogen bonds. These base pairs form the rungs of the ladder, and their precise pairing ensures the accuracy of genetic information.

The sugar and phosphate groups alternate along the outer edges of the DNA molecule, forming the backbone of the helix. The strands are antiparallel, meaning one runs in the 5' to 3' direction, while the other runs 3' to 5'. This arrangement is critical for DNA replication and the enzymes that interact with it.

DNA's Stability and Flexibility

DNA's structure is remarkably stable yet flexible, allowing it to store information while remaining accessible. The hydrogen bonds between base pairs stabilize the helix, while the flexibility of the sugar-phosphate backbone permits the DNA to coil tightly into chromosomes or uncoil for replication and transcription.

Chromosomes are highly compact structures in which DNA is wound around proteins called histones. In humans, DNA is divided into 23 pairs of chromosomes, with one set inherited from each parent. This organization allows DNA to fit within the nucleus while maintaining access to specific regions as needed.

Genes and Non-Coding DNA

Genes are segments of DNA that contain instructions for making proteins. A typical gene includes a regulatory region, which determines when and where the gene is active, and a coding region, which provides the blueprint for protein synthesis. In humans, genes are spread across the genome, comprising about 1-2% of the total DNA.

The rest of the DNA, often referred to as non-coding DNA, was once thought to be "junk." However, researchers have discovered that non-coding regions serve regulatory and structural purposes. They control gene expression, contribute to chromosome stability, and produce functional RNA molecules, such as ribosomal RNA (rRNA) and transfer RNA (tRNA).

DNA Replication: Copying the Blueprint

Before a cell divides, it must duplicate its DNA to ensure that each daughter cell receives an identical copy. DNA replication is a highly orchestrated process that begins at specific sites called origins of replication. Enzymes called helicases unwind the double helix, creating a replication fork.

Each strand of the DNA acts as a template for synthesizing a new complementary strand. DNA polymerase, the enzyme responsible for building the new strand, adds nucleotides one at a time, pairing A with T and C with G. The leading strand is synthesized continuously, while the lagging strand is built in short fragments called Okazaki fragments, which are later joined by DNA ligase.

This semi-conservative mechanism ensures high fidelity, with the new DNA molecule containing one original strand and one newly synthesized strand. Errors during replication are rare due to proofreading mechanisms of DNA polymerase, which correct mismatched base pairs. However, when mistakes occur, they can result in mutations that may lead to disease or evolutionary changes.

Transcription: From DNA to RNA

Transcription is the process by which DNA is used to create RNA, a single-stranded molecule that serves as an intermediary between the genetic code and protein synthesis. The enzyme RNA polymerase binds to a gene's promoter region, unwinding the DNA to expose the template strand.

RNA polymerase synthesizes a complementary RNA strand by pairing A with uracil (U) instead of thymine. The resulting molecule, messenger RNA (mRNA), carries the genetic instructions out of the nucleus and into the cytoplasm, where ribosomes translate it into proteins. Transcription is tightly regulated to ensure that genes are expressed at the right time and in the right amount.

Translation: Turning RNA Into Proteins

Translation occurs in the ribosome, a molecular machine composed of rRNA and proteins. The mRNA sequence is read in sets of three bases, called codons, each specifying a particular amino acid. For example, the codon AUG codes for methionine, the start signal for protein synthesis.

Transfer RNA (tRNA) molecules bring the appropriate amino acids to the ribosome, where they are linked together in the order dictated by the mRNA sequence. This process produces a polypeptide chain that folds into a functional protein. Proteins perform nearly every cellular task, from catalyzing reactions to providing structural support.

Regulation of DNA and Gene Expression

DNA does not function in isolation; its activity is finely tuned by cellular mechanisms. Regulatory sequences, such as enhancers and silencers, modulate gene expression by interacting with transcription factors. Epigenetic modifications, such as DNA methylation and histone acetylation, alter the accessibility of DNA without changing its sequence.

Environmental factors, including diet, stress, and toxins, can influence these regulatory systems, demonstrating how external conditions shape genetic activity. These interactions are part of why identical twins, who share the same DNA, can exhibit differences in health and behavior.

Mutations: Changes in the Blueprint

Mutations are permanent alterations in the DNA sequence. They can arise spontaneously during replication or be induced by environmental factors such as radiation, chemicals, or viruses. Point mutations, which change a single base pair, can have varying effects depending on their location. For example, a substitution in a coding region might alter an amino acid, potentially disrupting protein function.

Some mutations have minimal impact, while others can cause diseases like sickle cell anemia, which results from a single amino acid substitution in hemoglobin. Mutations in regulatory regions can also affect gene expression, leading to conditions such as cancer. On a broader scale, mutations drive evolution by introducing genetic variation.

DNA Repair Mechanisms

The stability of DNA is safeguarded by repair systems that detect and correct damage. For example, base excision repair removes chemically altered bases, while nucleotide excision repair fixes bulky distortions caused by UV light. Double-strand breaks, which are especially dangerous, are repaired through processes like homologous recombination or non-homologous end joining.

Defects in repair mechanisms can lead to genomic instability and increase susceptibility to diseases. Xeroderma pigmentosum, for instance, is a disorder caused by impaired nucleotide excision repair, making individuals highly sensitive to UV radiation and prone to skin cancer.

DNA in Forensics and Medicine

DNA's unique sequence makes it common in forensics, where it is used to identify individuals with extraordinary accuracy. Techniques like polymerase chain reaction (PCR) amplify DNA from minute samples, enabling its analysis in criminal investigations or paternity testing.

In medicine, DNA sequencing has revolutionized diagnostics and treatment. Genetic testing can identify mutations linked to inherited diseases, allowing for early interventions. Personalized medicine tailors treatments based on an individual's genetic makeup, optimizing efficacy and minimizing side effects.

DNA and Biotechnology

Biotechnology harnesses DNA to create products and solve problems. Recombinant DNA technology, which involves combining DNA from different sources, is used to produce insulin for diabetes treatment and create genetically modified crops. CRISPR-Cas9, a groundbreaking gene-editing tool, allows scientists to precisely modify DNA, offering potential cures for genetic disorders and advancing research in human biology.

Gene Expression: From DNA to Protein

Gene expression is the process by which information encoded in DNA is used to produce functional proteins. This multistep process involves transcription, RNA processing, and translation. Each step is tightly regulated, ensuring that proteins are produced at the right time, in the right place, and in the correct amounts to maintain cellular function.

Transcription: DNA to RNA

The first step in gene expression is transcription, where a specific segment of DNA is copied into RNA. This process occurs in the nucleus and is initiated when RNA polymerase binds to the promoter region of a gene. The promoter contains specific DNA sequences, such as the TATA box, which signal where transcription begins.

RNA polymerase unwinds the DNA double helix and synthesizes a complementary RNA strand using the DNA template strand. The RNA sequence is similar to the coding DNA strand but substitutes uracil (U) for thymine (T). For instance, a DNA template sequence of A-T-G-C would produce an RNA sequence of U-A-C-G.

Transcription produces a precursor molecule called pre-mRNA, which contains both coding regions (exons) and non-coding regions (introns). The inclusion of introns requires further processing before the RNA can be functional.

RNA Processing

Before leaving the nucleus, pre-mRNA undergoes splicing, where introns are removed, and exons are joined together to form mature mRNA. This process is carried out by the spliceosome, a complex of proteins and small nuclear RNAs (snRNAs). Splicing can occur in multiple ways, generating different protein products from a single gene. This phenomenon, known as alternative splicing, increases the diversity of proteins that a cell can produce.

In addition to splicing, the 5' end of the pre-mRNA is capped with a modified guanine nucleotide, and the 3' end is polyadenylated with a tail of adenine nucleotides. The 5' cap and poly-A tail protect the mRNA from degradation and facilitate its transport out of the nucleus.

Translation: RNA to Protein

Translation occurs in the cytoplasm at ribosomes, molecular machines composed of rRNA and proteins. The mature mRNA serves as a template for assembling a polypeptide chain. Translation begins at the start codon (AUG), which codes for methionine, and proceeds until a stop codon (UAA, UAG, or UGA) is encountered.

Transfer RNA (tRNA) molecules bring amino acids to the ribosome. Each tRNA has an anticodon that recognizes a specific codon on the mRNA. For example, the codon UUU, which codes for phenylalanine, pairs with a tRNA carrying the anticodon AAA. The ribosome links amino acids together through peptide bonds, forming a growing polypeptide chain.

The process occurs in three phases: initiation, elongation, and termination. During initiation, the ribosome assembles around the start codon. Elongation involves the sequential addition of amino acids, while termination releases the completed polypeptide when a stop codon is reached.

Protein Folding and Post-Translational Modifications

After translation, the polypeptide chain folds into its functional three-dimensional shape. This folding is guided by the sequence of amino acids and often assisted by chaperone proteins. Misfolding can lead to nonfunctional or toxic proteins, as seen in diseases like Alzheimer's.

Many proteins undergo post-translational modifications to become fully active. These modifications include phosphorylation, glycosylation, and cleavage of precursor molecules. For instance, insulin is synthesized as a single chain and cleaved into its active form after translation.

Regulation of Gene Expression

Cells regulate gene expression at multiple levels to respond to environmental cues, developmental signals, and internal demands. Transcriptional control is the primary regulatory mechanism. Transcription factors bind to regulatory DNA sequences, such as enhancers and silencers, to activate or repress gene transcription.

Epigenetic modifications, such as DNA methylation and histone acetylation, also influence gene expression by altering the accessibility of DNA to transcriptional machinery. For example, methylation of promoter regions typically silences gene activity.

Post-transcriptional regulation includes alternative splicing, mRNA stability, and translation efficiency. MicroRNAs (miRNAs) are small non-coding RNAs that bind to mRNA and prevent translation, fine-tuning protein production.

Inheritance Patterns and Genetic Disorders

Inheritance patterns describe how genetic traits and disorders are passed from one generation to the next. These patterns depend on the type of gene involved, its location, and whether the trait is dominant or recessive. Understanding these patterns provides insight into how genetic disorders arise and are transmitted.

Mendelian Inheritance

Gregor Mendel's principles of inheritance, established through experiments with pea plants, laid the foundation for understanding genetic traits. Mendelian inheritance includes autosomal dominant, autosomal recessive, X-linked dominant, and X-linked recessive patterns.

In **autosomal dominant inheritance**, a single copy of a mutated gene on a non-sex chromosome is sufficient to cause the trait or disorder. Individuals with an affected parent have a 50% chance of inheriting the condition. Examples include Huntington's disease, caused by mutations in the HTT gene, and Marfan syndrome, linked to mutations in the FBN1 gene.

Autosomal recessive inheritance requires two copies of the mutated gene for the trait or disorder to manifest. Carriers, with only one mutated gene, are typically unaffected. If both parents are carriers, their offspring have a 25% chance of inheriting the disorder. Cystic fibrosis and sickle cell anemia are autosomal recessive conditions.

X-linked inheritance involves genes located on the X chromosome. In X-linked dominant inheritance, one mutated copy of the gene causes the disorder, affecting both males and females. Rett syndrome is an example. In X-linked recessive inheritance, males are more frequently affected because they have only one X chromosome. Conditions like hemophilia and Duchenne muscular dystrophy follow this pattern.

Non-Mendelian Inheritance

Not all traits follow Mendelian patterns. Non-Mendelian inheritance includes incomplete dominance, codominance, mitochondrial inheritance, and polygenic traits.

In **incomplete dominance**, the heterozygous phenotype is an intermediate between the two homozygous phenotypes. For example, in familial hypercholesterolemia, individuals with one mutant allele have moderately elevated cholesterol levels, while those with two mutant alleles have severe elevations.

Codominance occurs when both alleles in a heterozygote are fully expressed. The ABO blood group system illustrates this, where individuals with IA and IB alleles have type AB blood, expressing both antigens.

Mitochondrial inheritance involves genes in the mitochondrial genome, which are passed exclusively from mother to offspring. Mitochondrial disorders, such as Leber hereditary optic neuropathy, affect energy production and typically involve tissues with high energy demands, like muscles and nerves.

Polygenic traits, like height and skin color, result from the interaction of multiple genes. Environmental factors also influence these traits, making their inheritance complex.

Chromosomal Disorders

Chromosomal abnormalities arise from changes in chromosome number or structure. **Aneuploidy**, the presence of an abnormal number of chromosomes, is a common cause of genetic disorders. For example, Down syndrome results from an extra copy of chromosome 21 (trisomy 21), while Turner syndrome arises from a missing X chromosome in females (45, X).

Structural changes, such as deletions, duplications, inversions, and translocations, can disrupt gene function. Cri du chat syndrome, caused by a deletion on chromosome 5, leads to developmental delays and distinct physical features.

Mutations and Genetic Disorders

Mutations, or changes in the DNA sequence, can cause genetic disorders by altering protein function. Point mutations, insertions, deletions, and frameshift mutations all have distinct effects. For instance, a single base substitution in the HBB gene causes sickle cell anemia, while a frameshift mutation in the CFTR gene leads to cystic fibrosis.

Some disorders result from dynamic mutations, where repetitive DNA sequences expand with each generation. Huntington's disease and fragile X syndrome are examples, with symptoms worsening as the mutation lengthens.

Multifactorial Disorders

Multifactorial disorders result from a combination of genetic and environmental factors. These include conditions like diabetes, heart disease, and certain cancers. While no single gene is responsible, specific genetic variants increase susceptibility. Lifestyle factors, such as diet, exercise, and exposure to toxins, interact with genetic predispositions to determine risk.

Genetic Counseling and Testing

Genetic counseling helps individuals and families understand inheritance patterns and assess the risk of passing on genetic conditions. Testing methods, such as karyotyping, gene sequencing, and genome-wide association studies, identify genetic mutations and chromosomal abnormalities. Early detection of genetic disorders allows for interventions, including prenatal therapies, lifestyle adjustments, and personalized medicine.

Modern Genetics: CRISPR and the Future of Gene Editing

CRISPR (Clustered Regularly Interspaced Short Palindromic Repeats) has transformed modern genetics by offering a precise and versatile tool for editing DNA. This groundbreaking technology allows scientists to modify genetic material with unprecedented accuracy, enabling applications in medicine, agriculture, and basic research. CRISPR's impact is rooted in its efficiency, simplicity, and ability to target specific DNA sequences.

Origins of CRISPR

The discovery of CRISPR began with observations in bacteria. Researchers found repeating DNA sequences in bacterial genomes interspersed with unique spacer sequences. These spacers were derived from viral DNA, serving as a genetic memory of past infections. When a virus attacked, bacteria used this system to recognize and destroy the invader. This immune response relies on two key components: CRISPR-associated proteins (Cas) and CRISPR RNA (crRNA).

The most widely used Cas protein, Cas9, is an endonuclease that cuts DNA at a specific site. The crRNA guides Cas9 to the target sequence, matching it through complementary base pairing. Scientists recognized the potential of this system for gene editing, adapting it for use in eukaryotic cells.

How CRISPR Works

CRISPR editing begins with the design of a guide RNA (gRNA), which directs the Cas9 protein to a specific DNA sequence. The gRNA contains a region complementary to the target sequence and a scaffold region that binds to Cas9. Once the gRNA-Cas9 complex locates the target, Cas9 creates a double-strand break in the DNA.

After the DNA is cut, the cell's repair mechanisms are triggered. These pathways can be exploited for editing purposes:

1. **Non-Homologous End Joining (NHEJ):** This repair process joins the broken DNA ends without a template. It often introduces small insertions or deletions (indels), disrupting the target gene's function. NHEJ is commonly used for creating gene knockouts.

2. **Homology-Directed Repair (HDR):** HDR uses a template to precisely repair the break. Scientists provide a synthetic DNA template containing the desired sequence, allowing for precise edits, such as correcting mutations or inserting new genes.

CRISPR's specificity depends on the 20-nucleotide sequence in the gRNA. However, the system requires a protospacer adjacent motif (PAM), a short DNA sequence near the target site, for Cas9 binding. This PAM requirement ensures proper targeting while reducing off-target effects.

Applications in Medicine

CRISPR has opened new possibilities for treating genetic diseases. By targeting disease-causing mutations, CRISPR offers potential cures for conditions previously considered untreatable.

In **sickle cell anemia**, CRISPR has been used to reactivate fetal hemoglobin production. Researchers edited the BCL11A gene, which suppresses fetal hemoglobin in adults, restoring a form of hemoglobin unaffected by the mutation. This approach is currently being tested in clinical trials, offering hope for millions of patients.

Cystic fibrosis, caused by mutations in the CFTR gene, is another target for CRISPR. Researchers have corrected CFTR mutations in cell cultures, paving the way for therapies that could restore lung function and reduce symptoms.

CRISPR also shows promise in treating **cancer**. Scientists are engineering immune cells, such as T-cells, to recognize and attack tumors. By editing genes that suppress immune responses, CRISPR enhances the effectiveness of immunotherapy. For example, CRISPR-edited CAR-T cells have been designed to target leukemia cells with greater precision.

Somatic vs. Germline Editing

Gene editing can be applied to somatic (non-reproductive) or germline (reproductive) cells. Somatic editing affects only the individual, making it less ethically contentious. Germline editing, on the other hand, introduces changes that are heritable, passing to future generations.

While germline editing offers the potential to eliminate genetic diseases at their source, it raises significant ethical concerns. Altering the germline could lead to unintended consequences, such as off-target effects or unforeseen impacts on development. Additionally, the prospect of "designer babies"—where traits like intelligence or physical appearance are edited—sparks fears of social inequality and eugenics.

In 2018, controversy erupted when a Chinese scientist announced the birth of CRISPR-edited twins. The embryos were modified to disable the CCR5 gene, potentially conferring resistance to HIV. The experiment was widely condemned for its lack of oversight, ethical violations, and disregard for safety.

Applications Beyond Medicine

CRISPR is revolutionizing agriculture by enabling precise modifications to crop genomes. Researchers have developed crops with increased resistance to pests, diseases, and environmental stress. For example, CRISPR-edited rice varieties tolerate drought and salinity better than traditional strains, addressing food security in regions affected by climate change.

In livestock, CRISPR has been used to create disease-resistant animals. Scientists have edited pigs to resist porcine reproductive and respiratory syndrome (PRRS), a devastating viral disease. These advances improve animal welfare and reduce the need for antibiotics.

CRISPR is also being applied to synthetic biology. By engineering bacteria and yeast, scientists can produce biofuels, pharmaceuticals, and biodegradable materials. For example, CRISPR-modified microbes are used to synthesize artemisinin, a key antimalarial drug, more efficiently than traditional methods.

Research and Functional Genomics

CRISPR has become an indispensable tool in basic research, allowing scientists to study gene function with precision. By knocking out or modifying specific genes, researchers can investigate their roles in development, disease, and physiology.

Functional genomics studies often use CRISPR-based screens to identify genes associated with particular traits. For instance, genome-wide CRISPR screens have identified genes essential for cancer cell survival, revealing potential drug targets.

Additionally, CRISPR is advancing our understanding of epigenetics. Scientists have developed CRISPR variants, such as dCas9 (dead Cas9), which bind to DNA without cutting it. These tools enable the study of gene regulation by altering chromatin structure or modifying DNA methylation.

Ethical and Regulatory Challenges

The rapid development of CRISPR technology has outpaced the establishment of ethical and regulatory frameworks. Balancing the benefits of gene editing with potential risks is a complex challenge.

One major concern is **off-target effects**, where CRISPR inadvertently edits unintended regions of the genome. While advances in gRNA design and Cas9 variants have improved specificity, the risk of unintended mutations remains. These off-target effects could have harmful consequences, particularly in clinical applications.

Another challenge is equitable access to CRISPR technology. High costs and technical expertise limit its availability, raising concerns about global disparities in healthcare and agriculture. Ensuring that CRISPR benefits are distributed fairly will require international cooperation and policy development.

Environmental concerns also arise in the context of gene drives, a CRISPR-based technique used to spread specific genetic traits through populations. Gene drives have been proposed to control invasive species or eradicate disease vectors like mosquitoes. However, their potential to disrupt ecosystems and spread beyond intended targets highlights the need for cautious implementation.

Future Directions

CRISPR technology continues to evolve, with new tools expanding its capabilities. Base editors, which can change a single nucleotide without cutting DNA, offer a gentler alternative to traditional CRISPR. For example, base editors have been used to correct point mutations in diseases like Duchenne muscular dystrophy.

Prime editing, a newer technique, allows precise DNA modifications without relying on double-strand breaks or donor templates. This method has shown promise in correcting a wider range of mutations, increasing the scope of CRISPR applications.

Advances in delivery methods are also improving CRISPR's effectiveness. Researchers are developing lipid nanoparticles, viral vectors, and electroporation techniques to deliver CRISPR components more efficiently to target cells.

CRISPR's integration with artificial intelligence (AI) and bioinformatics is streamlining guide RNA design and predicting off-target effects. These innovations enhance the precision and safety of gene editing, accelerating its adoption in clinical and industrial settings.

The development of ethical guidelines and international collaboration will be critical as CRISPR reshapes genetics and biotechnology. While challenges remain, the potential of CRISPR to address genetic diseases, improve food security, and advance scientific knowledge is unmatched, signaling a new era in DNA and genetics.

CHAPTER 4: THE CELL - LIFE'S FUNDAMENTAL UNIT

Cellular Structures and Their Functions: Organelles and Membranes

The cell is the fundamental unit of life, with a complex internal structure that allows it to carry out the myriad functions necessary for survival and growth. Organelles, each with specialized roles, and the cell membrane, which acts as a selective barrier, form the intricate architecture of a cell. Together, they ensure efficient functioning and adaptability.

The Cell Membrane: The Gatekeeper

The cell membrane, also called the plasma membrane, surrounds the cell, providing a dynamic boundary between the internal environment and the outside world. Composed of a **phospholipid bilayer**, it contains hydrophilic heads facing outward and hydrophobic tails facing inward. This arrangement creates a semi-permeable barrier that regulates the movement of substances in and out of the cell.

Embedded within the bilayer are proteins that serve various functions, such as transport, signaling, and structural support. **Channel proteins** allow ions like sodium and potassium to move across the membrane, while **carrier proteins** facilitate the transport of larger molecules like glucose. Receptor proteins on the surface detect chemical signals, such as hormones, triggering cellular responses.

The membrane is fluid, with molecules constantly shifting within it. Cholesterol molecules stabilize its structure, preventing it from becoming too rigid or too fluid. Attached carbohydrates form glycoproteins and glycolipids, which are critical for cell recognition and communication.

The Nucleus: The Command Center

The nucleus houses the cell's genetic material and coordinates activities such as growth, metabolism, and protein synthesis. Enclosed by the **nuclear envelope**, a double membrane with nuclear pores, it controls access to DNA. These pores regulate the exchange of materials, allowing molecules like RNA and proteins to pass in and out while safeguarding the genetic information.

Inside the nucleus, **chromatin**, a combination of DNA and proteins, is organized into chromosomes during cell division. The nucleolus, a dense region within the nucleus, is the site of **ribosome assembly**. Ribosomal RNA (rRNA) is synthesized here and combined with proteins to form ribosomal subunits, which are then exported to the cytoplasm.

Mitochondria: The Powerhouses

Mitochondria generate ATP, the energy currency of the cell, through aerobic respiration. These double-membraned organelles have an outer membrane that acts as a barrier and an inner membrane folded into **cristae**, which increase the surface area for biochemical reactions. The matrix, the space enclosed by the inner membrane, contains enzymes, mitochondrial DNA, and ribosomes.

During cellular respiration, glucose is broken down into carbon dioxide and water, releasing energy stored in chemical bonds. This energy is used to produce ATP in the electron transport chain, which is located in the inner membrane. Mitochondria are unique because they have their own DNA, allowing them to replicate independently of the cell.

Endoplasmic Reticulum: The Manufacturing Center

The **endoplasmic reticulum (ER)** is a network of membranes extending from the nuclear envelope, involved in the synthesis, folding, and transport of proteins and lipids. It comes in two forms: rough ER and smooth ER.

The **rough ER** is studded with ribosomes, giving it a grainy appearance. Ribosomes on the rough ER synthesize proteins destined for secretion, insertion into the membrane, or delivery to organelles. The ER modifies these proteins by folding them into their functional shapes and adding carbohydrate groups.

The **smooth ER**, lacking ribosomes, is involved in lipid synthesis, detoxification of harmful substances, and calcium ion storage. In liver cells, the smooth ER detoxifies drugs and alcohol, while in muscle cells, it regulates calcium release during contraction.

Ribosomes: The Protein Factories

Ribosomes are small, dense structures composed of rRNA and proteins. Found either floating freely in the cytoplasm or attached to the rough ER, they are the sites of protein synthesis. Free ribosomes produce proteins used within the cell, such as enzymes for metabolism, while bound ribosomes produce proteins for export or for use in membranes and organelles.

Ribosomes translate the genetic code carried by messenger RNA (mRNA) into amino acid sequences, assembling them into polypeptides. These polypeptides are then folded and modified to become functional proteins.

Golgi Apparatus: The Packaging and Distribution Hub

The **Golgi apparatus** is a stack of flattened membranes responsible for processing, packaging, and distributing proteins and lipids. Proteins synthesized in the ER are transported to the Golgi in vesicles. Once there, they are modified—such as by adding sugar molecules to form glycoproteins—and sorted based on their destination.

The Golgi packages these molecules into vesicles for delivery. For example, secretory vesicles carry proteins to the cell membrane for release, while lysosomes, another type of vesicle, contain enzymes for digestion. The Golgi's role in post-translational modification ensures that proteins and lipids are functional and properly directed.

Lysosomes: The Recycling Centers

Lysosomes are membrane-bound organelles containing digestive enzymes that break down macromolecules, old organelles, and foreign substances. These enzymes function best in acidic conditions, maintained by proton pumps within the lysosomal membrane. When cellular components become damaged or obsolete, lysosomes fuse with them, degrading them into reusable molecules.

Lysosomes are essential for cellular cleanup and defense. In immune cells, they help destroy pathogens engulfed during phagocytosis. Dysfunctional lysosomes are associated with disorders like Tay-Sachs disease, where undigested materials accumulate, impairing cellular function.

Peroxisomes: The Detox Units

Peroxisomes are small, membrane-bound organelles involved in detoxification and lipid metabolism. They contain enzymes like catalase, which breaks down hydrogen peroxide, a byproduct of metabolic reactions, into water and oxygen. Peroxisomes also metabolize fatty acids and synthesize plasmalogens, important components of myelin in nerve cells.

These organelles are particularly abundant in liver and kidney cells, where they neutralize harmful substances and support energy production by breaking down long-chain fatty acids.

Cytoskeleton: The Cellular Scaffold

The cytoskeleton is a dynamic network of protein filaments that provides structural support, facilitates movement, and organizes cellular contents. It consists of three main components: microfilaments, intermediate filaments, and microtubules.

- **Microfilaments**, composed of actin, are involved in cell shape, movement, and division. They form the contractile ring during cytokinesis and drive cellular crawling in processes like wound healing.
- **Intermediate filaments** provide mechanical strength, maintaining the cell's shape and anchoring organelles in place. Keratin is an example found in skin cells.

- **Microtubules**, made of tubulin, form hollow rods that act as tracks for intracellular transport. Motor proteins like kinesin and dynein move vesicles and organelles along microtubules. Microtubules also compose the spindle fibers during cell division and are the structural elements of cilia and flagella.

Vacuoles and Vesicles: Storage and Transport

In human cells, vesicles and vacuoles serve as storage and transport structures. Vesicles shuttle molecules between organelles, while vacuoles, though larger and more prominent in plant cells, exist in human cells as smaller compartments for storage.

Specialized vesicles like secretory vesicles and endosomes support processes like exocytosis and endocytosis, enabling cells to communicate with their environment and regulate internal conditions.

Centrosomes and Centrioles: Organizing Microtubules

The centrosome is the microtubule-organizing center of the cell, located near the nucleus. It contains a pair of centrioles, cylindrical structures made of microtubules. During cell division, centrosomes orchestrate the formation of the mitotic spindle, which separates chromosomes into daughter cells.

Centrioles also contribute to the assembly of cilia and flagella, enabling cell movement and fluid transport across surfaces.

Endomembrane System: Coordinating Activities

The endomembrane system includes the ER, Golgi apparatus, lysosomes, vesicles, and the cell membrane. These interconnected structures collaborate to synthesize, modify, transport, and recycle cellular materials. For instance, a protein synthesized in the rough ER may be modified in the Golgi, transported in a vesicle, and secreted via exocytosis.

This coordination ensures efficiency, allowing the cell to adapt to changing demands and maintain homeostasis.

Cellular Respiration and Energy Production (ATP)

Cellular respiration is the process by which cells extract energy from nutrients and convert it into ATP, the primary molecule used for energy transfer. This multi-step process occurs in the cytoplasm and mitochondria, breaking down glucose and other fuels into carbon dioxide and water, releasing energy in a controlled manner.

Overview of Cellular Respiration

Cellular respiration involves three main stages: glycolysis, the citric acid cycle (Krebs cycle), and oxidative phosphorylation. Each stage contributes to the production of ATP while generating intermediates used in subsequent steps.

1. **Glycolysis** occurs in the cytoplasm and does not require oxygen. One glucose molecule (6 carbons) is split into two pyruvate molecules (3 carbons each), yielding 2 ATP molecules and 2 NADH molecules. NADH is a carrier that transfers electrons to the mitochondria for further energy production. Glycolysis is crucial because it provides energy even in anaerobic conditions.

2. **The Citric Acid Cycle** takes place in the mitochondrial matrix. Pyruvate is first converted to acetyl-CoA, which enters the cycle. Each turn of the cycle produces 3 NADH, 1 FADH2, and 1 ATP, along with carbon dioxide as a byproduct. For each glucose molecule, the cycle runs twice, generating additional electron carriers and metabolic intermediates.

3. **Oxidative Phosphorylation** occurs across the inner mitochondrial membrane. NADH and FADH2 donate electrons to the electron transport chain (ETC), a series of protein complexes embedded in the membrane. As electrons pass through the chain, protons are pumped into the intermembrane space, creating an electrochemical gradient. This gradient drives ATP synthesis through ATP synthase, a molecular turbine that converts ADP and inorganic phosphate into ATP. This stage produces the majority of ATP, up to 34 molecules per glucose.

Aerobic vs. Anaerobic Respiration

Aerobic respiration requires oxygen, which acts as the final electron acceptor in the ETC, forming water. In the absence of oxygen, cells rely on anaerobic pathways, such as fermentation. In human cells, anaerobic respiration converts pyruvate to lactate, regenerating NAD+ to sustain glycolysis. However, anaerobic respiration yields only 2 ATP per glucose, compared to 36-38 ATP in aerobic conditions.

Energy Yield and Efficiency

The total ATP yield from one glucose molecule includes 2 ATP from glycolysis, 2 ATP from the citric acid cycle, and up to 34 ATP from oxidative phosphorylation, totaling 36-38 ATP. The variability depends on the efficiency of NADH transport into the mitochondria. This process is remarkably efficient, with approximately 40% of the energy in glucose captured as ATP; the rest is released as heat.

Role of Other Fuels

While glucose is the primary fuel for cellular respiration, other macromolecules can also provide energy. Fatty acids undergo beta-oxidation in the mitochondria, generating acetyl-CoA, which enters the citric acid cycle. Fats are a highly efficient energy source, yielding more ATP per molecule than carbohydrates. Proteins, when used for energy, are broken down into amino acids, which are deaminated and converted into intermediates that enter the respiration pathway.

Importance of ATP

ATP serves as the universal energy currency in cells. It powers muscle contraction, active transport across membranes, biosynthesis of macromolecules, and signal transduction. The continuous production of ATP through cellular respiration is essential for maintaining cellular functions and homeostasis.

The Cell Cycle and Mitosis

The cell cycle is a series of events that cells go through to grow, replicate their DNA, and divide into two identical daughter cells. This process is essential for growth, tissue repair, and reproduction in multicellular organisms. The cycle consists of interphase and the mitotic phase, each with distinct stages.

Interphase: Preparation for Division

Interphase is the longest phase of the cell cycle, during which the cell grows, replicates its DNA, and prepares for division. It is divided into three stages: G1, S, and G2.

1. **G1 Phase (Gap 1):** During G1, the cell grows in size, synthesizes proteins, and produces organelles. This phase is critical for ensuring the cell is ready to enter the S phase. Cells also monitor their environment and check for DNA damage. If conditions are unfavorable, cells may enter G0, a quiescent state where they cease to divide.

2. **S Phase (Synthesis):** In the S phase, the cell replicates its DNA, ensuring that each daughter cell receives an identical copy. Each chromosome is duplicated to form two sister chromatids held together by a centromere. The cell also duplicates its centrosomes, which will organize the mitotic spindle during division.

3. **G2 Phase (Gap 2):** In G2, the cell continues to grow and synthesizes proteins needed for mitosis, such as microtubule components. The cell checks for DNA replication errors and repairs them to ensure genomic integrity.

Mitosis: Nuclear Division

Mitosis is the process of dividing the replicated chromosomes into two nuclei, ensuring that each daughter cell receives an identical set of genetic material. It is divided into five stages: prophase, prometaphase, metaphase, anaphase, and telophase.

1. **Prophase:** Chromosomes condense into visible structures, and each consists of two sister chromatids. The nuclear envelope begins to break down, and the mitotic spindle, composed of microtubules, forms. Centrosomes migrate to opposite poles of the cell.

2. **Prometaphase:** The nuclear envelope fully disintegrates, allowing spindle fibers to attach to the centromeres of chromosomes at specialized protein complexes called kinetochores. Chromosomes begin to move toward the center of the cell.

3. **Metaphase:** Chromosomes align at the metaphase plate, an imaginary line equidistant from the two spindle poles. This alignment ensures that each daughter cell will receive one copy of each chromosome.

4. **Anaphase:** Sister chromatids are pulled apart as spindle fibers shorten, moving them toward opposite poles of the cell. This separation ensures that each pole receives an identical set of chromosomes.

5. **Telophase:** Chromosomes reach the poles and begin to decondense. The nuclear envelope re-forms around each set of chromosomes, creating two separate nuclei. The spindle apparatus disassembles.

Cytokinesis: Cytoplasmic Division

Cytokinesis occurs after mitosis and involves the division of the cytoplasm to form two distinct daughter cells. In animal cells, a contractile ring of actin and myosin filaments forms around the equator of the cell, creating a cleavage furrow that pinches the cell in two. In plant cells, a cell plate forms along the center, which develops into a new cell wall separating the daughter cells.

Regulation of the Cell Cycle

The cell cycle is tightly regulated by checkpoints that ensure each phase is completed correctly before the next begins. The three main checkpoints are:

1. **G1 Checkpoint:** Assesses cell size, nutrient availability, and DNA integrity. If conditions are unsuitable, the cell may enter G0 or undergo apoptosis.
2. **G2 Checkpoint:** Verifies that DNA replication is complete and error-free. If errors are detected, the cell cycle is paused for repair.

3. **M Checkpoint (Spindle Checkpoint):** Ensures that all chromosomes are properly attached to spindle fibers before progressing to anaphase.

These checkpoints are controlled by cyclins and cyclin-dependent kinases (CDKs). Cyclins are proteins whose levels fluctuate during the cell cycle, while CDKs are enzymes activated by cyclins. Together, they regulate the progression of the cell cycle.

Disruptions in the Cell Cycle

Abnormal regulation of the cell cycle can lead to uncontrolled cell division, a hallmark of cancer. Mutations in genes like p53, a tumor suppressor that halts the cycle for DNA repair, allow damaged cells to proliferate. Understanding the mechanisms of cell cycle regulation has led to targeted therapies, such as drugs that inhibit CDKs to slow cancer progression.

Apoptosis: The Science of Programmed Cell Death

Apoptosis is a highly regulated process of programmed cell death essential for maintaining the health and balance of tissues in multicellular organisms. Unlike necrosis, which is a chaotic form of cell death resulting from injury or infection, apoptosis is controlled, predictable, and beneficial. It ensures the removal of damaged, unnecessary, or potentially harmful cells without triggering inflammation.

Key Features of Apoptosis

Apoptosis involves a series of distinct morphological and biochemical changes. Cells undergoing apoptosis shrink and lose their attachment to neighboring cells. The plasma membrane remains intact but forms blebs—small bulges indicating cytoskeletal changes. Inside the cell, the chromatin condenses and fragments, and the nucleus breaks into smaller pieces. Cellular components are packaged into membrane-bound vesicles called **apoptotic bodies,** which are quickly engulfed and digested by phagocytic cells. This prevents leakage of cellular contents and protects surrounding tissues from damage.

The Role of Apoptosis in Development and Homeostasis

During embryonic development, apoptosis shapes organs and tissues by eliminating unnecessary cells. For example, the separation of fingers and toes requires the apoptosis of cells in the tissue between them. In the nervous system, apoptosis removes excess neurons, ensuring that only those making proper connections survive.

In adults, apoptosis maintains tissue homeostasis by balancing cell proliferation and death. It eliminates cells with irreparable DNA damage, thus preventing mutations from propagating. The immune system uses apoptosis to remove self-reactive lymphocytes that could attack the body's own tissues, reducing the risk of autoimmune diseases. Apoptosis also has a role in removing cells infected by viruses or affected by stress, ensuring the integrity of the organism.

Molecular Mechanisms of Apoptosis

Apoptosis is orchestrated by a family of proteases called **caspases**, which exist as inactive precursors (procaspases) and are activated in response to apoptotic signals. The process is divided into two main pathways: the intrinsic (mitochondrial) pathway and the extrinsic (death receptor) pathway. Both pathways converge on a common execution phase where caspases dismantle the cell.

1. **Intrinsic Pathway:** This pathway is triggered by internal stress signals, such as DNA damage, oxidative stress, or nutrient deprivation. The mitochondria have a central role by releasing cytochrome c into the cytoplasm. Cytochrome c binds to Apaf-1 (apoptotic protease activating factor-1), forming the apoptosome, which activates caspase-9. Caspase-9, in turn, activates executioner caspases like caspase-3 and caspase-7, leading to cell death. The intrinsic pathway is tightly regulated by the Bcl-2 family of proteins, which includes both pro-apoptotic (e.g., Bax, Bak) and anti-apoptotic (e.g., Bcl-2, Bcl-xL) members.

2. **Extrinsic Pathway:** The extrinsic pathway is initiated by external signals binding to death receptors on the cell surface. These receptors, such as Fas and TNF receptor, belong to the tumor necrosis factor (TNF) receptor family. When a ligand binds to these receptors, it recruits adaptor proteins like FADD (Fas-associated death domain) to form a death-inducing signaling complex (DISC). DISC activates caspase-8, which can directly activate executioner caspases or amplify the signal through the intrinsic pathway by cleaving Bid, a pro-apoptotic Bcl-2 protein.

The execution phase is common to both pathways and involves the cleavage of cellular proteins and DNA. Caspases target structural proteins, enzymes, and repair machinery, ensuring a controlled dismantling of the cell.

Regulation of Apoptosis

The decision to undergo apoptosis is tightly regulated to prevent premature or excessive cell death. The balance between pro-apoptotic and anti-apoptotic signals determines the cell's fate. Survival signals, such as growth factors, activate pathways like PI3K/Akt, which inhibit pro-apoptotic proteins and promote cell survival. Conversely, stress signals activate pathways that tip the balance toward apoptosis.

Proteins like p53, often called the "guardian of the genome," are important in regulating apoptosis in response to DNA damage. If damage is too severe to repair, p53 promotes the expression of pro-apoptotic genes, pushing the cell toward apoptosis. Dysregulation of p53 is a hallmark of many cancers, allowing cells with damaged DNA to escape apoptosis and proliferate uncontrollably.

Apoptosis in Disease

Apoptosis is essential for health, but its dysregulation is linked to numerous diseases. Insufficient apoptosis can result in uncontrolled cell growth, as seen in cancer. Many tumors have mutations in apoptotic pathways, such as overexpression of anti-apoptotic Bcl-2 proteins or loss of p53 function, enabling them to evade cell death. Therapies targeting these pathways, such as BH3 mimetics that inhibit Bcl-2, are being developed to restore apoptosis in cancer cells.

Excessive apoptosis contributes to degenerative diseases. In neurodegenerative conditions like Alzheimer's, Parkinson's, and Huntington's diseases, excessive neuronal apoptosis leads to the progressive loss of brain function. Similarly, in ischemic conditions such as heart attacks and strokes, apoptosis of cardiac and neuronal cells exacerbates tissue damage. Research into inhibitors of caspases and mitochondrial pathways aims to mitigate this excessive cell death.

Autoimmune diseases can result from defective apoptosis of self-reactive immune cells. When these cells fail to undergo apoptosis, they persist and attack healthy tissues. Systemic lupus erythematosus (SLE) is one example where impaired clearance of apoptotic cells leads to the accumulation of cellular debris and a heightened immune response.

Therapeutic Implications of Apoptosis

Understanding the mechanisms of apoptosis has opened avenues for therapeutic interventions. In cancer, strategies aim to enhance apoptosis in tumor cells. BH3 mimetics, which mimic pro-apoptotic proteins, overcome the resistance of cancer cells to apoptosis by inhibiting anti-apoptotic Bcl-2 proteins. Another approach involves activating death receptors with agonists like TRAIL (TNF-related apoptosis-inducing ligand), selectively inducing apoptosis in tumor cells while sparing normal cells.

In diseases of excessive apoptosis, such as neurodegeneration, researchers are exploring caspase inhibitors to protect cells. Experimental therapies aim to block the apoptotic cascade at key points, preserving cell viability. These therapies hold promise for reducing tissue damage in stroke, heart attack, and autoimmune diseases.

Gene therapy offers another potential avenue for regulating apoptosis. By introducing genes that restore p53 function or suppress overactive apoptotic pathways, scientists hope to correct imbalances in apoptosis that underlie various diseases.

Apoptosis and Immunity

Apoptosis is integral to the immune system, ensuring its proper function. During development, apoptosis eliminates self-reactive T-cells in the thymus, preventing autoimmunity. In response to infection, apoptosis clears infected cells, limiting the spread of pathogens. Cytotoxic T-cells and natural killer (NK) cells induce apoptosis in target cells by releasing perforin and granzyme, which activate the caspase cascade.

Phagocytes, such as macrophages, are critical in clearing apoptotic cells. This process, called efferocytosis, prevents the release of inflammatory signals. Apoptotic cells display "eat-me" signals, like phosphatidylserine, on their surface, which are recognized by phagocytes. Efficient clearance of apoptotic cells is essential for resolving inflammation and promoting tissue repair.

The precision and coordination of apoptosis highlight its importance as a cellular safeguard. By regulating cell death, apoptosis ensures the survival of the organism as a whole, maintaining the delicate balance between growth, repair, and elimination.

CHAPTER 5: TISSUES AND THEIR FUNCTIONS

Classification of Tissues: Epithelial, Connective, Muscle, and Nervous

Tissues are groups of cells that work together to perform specific functions. In human biology, tissues are classified into four main types: epithelial, connective, muscle, and nervous. Each type has distinct structures and roles that contribute to the overall function of organs and systems.

Epithelial Tissue: Covering and Protection

Epithelial tissue covers the body's surfaces, lines cavities, and forms glands. It acts as a barrier, regulates substance exchange, and provides sensory functions. Epithelial cells are tightly packed, forming continuous sheets with minimal extracellular matrix. These cells are connected by junctions like desmosomes and tight junctions, ensuring structural integrity and selective permeability.

Epithelial tissue is classified based on the number of cell layers and the shape of the cells:

1. **Simple Epithelium:** Composed of a single cell layer, it allows for efficient exchange of materials.

    - **Simple squamous epithelium** consists of flat cells, found in the lining of blood vessels (endothelium) and alveoli of the lungs, facilitating diffusion and filtration.
    - **Simple cuboidal epithelium**, with cube-shaped cells, is present in glands and kidney tubules, aiding in secretion and absorption.
    - **Simple columnar epithelium**, composed of tall cells, lines the digestive tract. Some cells have microvilli to increase surface area for absorption, while others secrete mucus.

2. **Stratified Epithelium:** Made up of multiple layers, it provides protection against physical and chemical stress.

    - **Stratified squamous epithelium** is found in the skin, mouth, and esophagus. The outer layer of the skin is keratinized, making it waterproof and durable.
    - **Stratified cuboidal and columnar epithelium** are less common and found in ducts of large glands.

3. **Pseudostratified Epithelium:** Appears layered due to nuclei at different levels but is actually a single layer. Pseudostratified columnar epithelium, often ciliated, lines the respiratory tract, where it traps and moves particles out of the airways.

4. **Transitional Epithelium:** Specialized for stretching, it is found in the urinary bladder. The cells change shape from round to flat as the bladder fills.

Epithelial tissue also forms glands:

- **Exocrine glands** release their products, such as enzymes or sweat, through ducts.
- **Endocrine glands** secrete hormones directly into the bloodstream.

Connective Tissue: Support and Binding

Connective tissue is the most diverse tissue type, providing structural support, binding organs, and transporting substances. It has fewer cells compared to epithelial tissue, embedded in an abundant extracellular matrix composed of fibers (collagen, elastin, and reticular) and ground substance.

1. **Loose Connective Tissue:** Characterized by loosely arranged fibers, it provides support and flexibility.
    - **Areolar tissue** cushions organs and contains fibroblasts, macrophages, and immune cells. It is found beneath epithelial layers and around blood vessels.
    - **Adipose tissue**, made of fat-storing adipocytes, insulates the body, stores energy, and cushions organs.
    - **Reticular tissue** forms a supportive framework for lymphoid organs like the spleen and lymph nodes.

2. **Dense Connective Tissue:** Contains tightly packed collagen fibers, offering strength and resistance to stretching.
    - **Dense regular connective tissue**, with parallel fibers, forms tendons (connecting muscles to bones) and ligaments (connecting bones to bones).
    - **Dense irregular connective tissue**, with fibers arranged in multiple directions, provides durability in the dermis of the skin and joint capsules.

3. **Cartilage:** A resilient and flexible connective tissue, cartilage lacks blood vessels and relies on diffusion for nutrient delivery.

- - **Hyaline cartilage**, found in the nose, trachea, and ends of long bones, provides smooth surfaces for joint movement.
  - **Elastic cartilage**, in the ear and epiglottis, offers flexibility and maintains shape.
  - **Fibrocartilage**, located in intervertebral discs and knee joints, absorbs shock due to its high collagen content.

4. **Bone (Osseous Tissue):** Bone tissue provides structural support and protection. It has a mineralized matrix containing calcium phosphate, making it rigid. Osteocytes, housed in lacunae, maintain the matrix, while osteoblasts build it and osteoclasts break it down for remodeling.

5. **Blood:** As a fluid connective tissue, blood transports gases, nutrients, hormones, and waste. Its cellular components include red blood cells (oxygen transport), white blood cells (immune defense), and platelets (clotting). Plasma, the liquid matrix, carries dissolved substances.

Muscle Tissue: Movement and Force

Muscle tissue generates force and movement by contracting in response to stimulation. It contains specialized proteins, actin and myosin, which slide past each other to produce contraction. Muscle tissue is classified into three types:

1. **Skeletal Muscle:** Skeletal muscle is attached to bones and enables voluntary movements. Its cells are long, cylindrical, multinucleated, and striated due to the organized arrangement of actin and myosin filaments. Skeletal muscle contracts rapidly and fatigues easily but is capable of powerful movements.

2. **Cardiac Muscle:** Found exclusively in the walls of the heart, cardiac muscle is responsible for pumping blood. Its cells are striated, branched, and connected by **intercalated discs**, which contain gap junctions and desmosomes. These structures ensure synchronized contraction and strong adhesion between cells. Cardiac muscle is involuntary and does not fatigue under normal conditions.

3. **Smooth Muscle:** Smooth muscle is found in the walls of hollow organs, such as the intestines, blood vessels, and bladder. Its cells are spindle-shaped, uninucleated, and non-striated. Smooth muscle contractions are involuntary and slower than skeletal muscle contractions but can be sustained for longer periods, as seen in the digestive process or vascular tone.

Nervous Tissue: Communication and Control

Nervous tissue forms the communication network of the body, transmitting signals between different regions to coordinate functions. It is composed of two main types of cells: neurons and glial cells.

1. **Neurons:** Neurons are the functional units of the nervous system. They have three main parts:
    - **Cell body (soma):** Contains the nucleus and organelles.
    - **Dendrites:** Short, branching processes that receive signals from other cells.
    - **Axon:** A long projection that transmits signals to other neurons, muscles, or glands. The axon is often covered by a **myelin sheath**, which insulates it and speeds up signal conduction. Gaps in the myelin sheath, called nodes of Ranvier, facilitate rapid signal transmission through saltatory conduction.

Neurons communicate through electrical impulses and chemical signals, transmitting information across synapses using neurotransmitters.

2. **Glial Cells:** Glial cells support, protect, and nourish neurons. They outnumber neurons and perform various functions:
    - **Astrocytes** regulate the extracellular environment and maintain the blood-brain barrier.
    - **Microglia** act as immune cells, removing debris and pathogens.
    - **Oligodendrocytes** in the central nervous system (CNS) and **Schwann cells** in the peripheral nervous system (PNS) produce myelin.

Nervous tissue forms the brain, spinal cord, and peripheral nerves, enabling sensory input, motor output, and integration of information.

Each type of tissue—epithelial, connective, muscle, and nervous—contributes to the structure and function of the human body in distinct ways. Their unique characteristics allow them to work together, forming the organs and systems that sustain life.

Specialized Functions of Each Tissue Type

Tissues are specialized for specific functions based on their structure and cellular composition. The four main tissue types—epithelial, connective, muscle, and nervous—work in unique ways to support the body's needs.

Epithelial Tissue: Selective Barriers and Secretory Functions

Epithelial tissue is specialized for covering surfaces, lining cavities, and forming glands. Its tightly packed cells and minimal extracellular matrix make it an effective barrier against physical damage, pathogens, and dehydration. **Epithelial cells act as gatekeepers**, regulating the movement of substances through diffusion, active transport, or secretion.

- In the **digestive tract**, columnar epithelium absorbs nutrients and secretes digestive enzymes and mucus. Specialized cells like goblet cells produce mucus to protect the lining from mechanical and chemical stress.
- In the **respiratory system**, pseudostratified columnar epithelium with cilia traps and moves debris-laden mucus out of the airways.
- Glandular epithelium forms **exocrine glands**, which secrete substances like sweat, saliva, or enzymes, and **endocrine glands**, which release hormones into the bloodstream. Hormones regulate distant processes such as growth, metabolism, and reproduction.

The protective, absorptive, and secretory roles of epithelial tissue make it essential for maintaining homeostasis and interacting with the environment.

Connective Tissue: Structural Support and Transport

Connective tissue connects, supports, and anchors other tissues and organs. Its cells are embedded in an extracellular matrix composed of fibers (collagen, elastin, and reticular fibers) and ground substance, giving it versatility and strength.

- **Bone tissue** provides structural support, protects internal organs, and serves as a reservoir for calcium and phosphate. Osteocytes maintain the mineralized matrix, while osteoblasts and osteoclasts regulate bone formation and resorption.
- **Adipose tissue** specializes in energy storage, insulation, and cushioning. Fat cells store triglycerides, releasing energy when needed. In addition to its metabolic role, adipose tissue helps regulate hormones like leptin, which influence appetite.
- **Cartilage** offers flexibility and shock absorption in joints and other structures. Hyaline cartilage reduces friction in articulating joints, elastic cartilage maintains shape in structures like the ear, and fibrocartilage absorbs impact in intervertebral discs.
- **Blood**, a fluid connective tissue, transports oxygen, nutrients, hormones, and waste products. Red blood cells deliver oxygen, white blood cells defend against infection, and platelets facilitate clotting.

Connective tissue adapts to provide structural integrity, protection, and the transport of essential substances.

Muscle Tissue: Force Generation and Movement

Muscle tissue is specialized for contraction, enabling movement, posture, and the circulation of blood. The contractile proteins **actin and myosin** are integral to its function.

- **Skeletal muscle** produces voluntary movements. It generates powerful contractions to move bones and support the body. Skeletal muscle cells, or fibers, are long, multinucleated, and striated. They can rapidly respond to neural signals, making them ideal for activities requiring precision and strength.
- **Cardiac muscle** pumps blood throughout the body. Its rhythmic, involuntary contractions are driven by specialized pacemaker cells in the heart. The presence of intercalated discs ensures that cardiac cells contract in unison, maintaining a steady and efficient heartbeat.
- **Smooth muscle** enables involuntary movements in internal organs. Its slow and sustained contractions are critical for processes like peristalsis in the digestive tract, blood vessel dilation and constriction, and bladder control.

Muscle tissue's ability to generate force and adapt to various functional demands ensures mobility, circulation, and vital physiological processes.

Nervous Tissue: Communication and Coordination

Nervous tissue specializes in transmitting signals to coordinate body functions. It relies on **neurons** to send electrical impulses and **glial cells** to provide support and maintenance.

- **Sensory neurons** detect stimuli from the environment, such as temperature, pressure, or light, and relay this information to the central nervous system.
- **Motor neurons** transmit signals from the brain and spinal cord to muscles and glands, enabling movement and responses.
- **Interneurons**, found in the brain and spinal cord, integrate information from sensory inputs and coordinate appropriate outputs.

Glial cells enhance nervous tissue function by maintaining the environment, insulating axons, and protecting against pathogens. Together, these components ensure precise communication and regulation of the body's systems.

How Tissues Work Together to Form Organs

Organs are formed by multiple tissue types working in unison to perform specific functions. Each tissue contributes unique properties, allowing the organ to execute complex tasks efficiently.

The Stomach: Digestion and Absorption

The stomach, an organ essential for digestion, integrates all four tissue types to process food and absorb nutrients. Its inner lining is made of **epithelial tissue**, which secretes mucus, enzymes, and hydrochloric acid. The mucus protects the stomach lining from the acidic environment, while the enzymes and acid break down food into a semi-liquid mixture called chyme.

Beneath the epithelial layer, **connective tissue** provides structural support and contains blood vessels to transport nutrients. The stomach's walls consist of multiple layers of **smooth muscle**, which contract rhythmically to churn food and mix it with digestive juices. **Nervous tissue** regulates these contractions and coordinates secretions, responding to signals like the presence of food or the stretching of the stomach.

The Heart: Circulation and Pumping Blood

The heart relies on the coordinated function of tissues to pump blood throughout the body. **Cardiac muscle** forms the bulk of the heart's walls, contracting rhythmically to propel blood. The muscle's structure, with intercalated discs, ensures synchronized contractions for effective pumping.

The inner lining of the heart, the **endocardium**, is composed of **epithelial tissue** that reduces friction as blood flows through the chambers. **Connective tissue** reinforces the heart's structure, forming the fibrous skeleton that supports valves and anchors muscle fibers. **Nervous tissue**, including the sinoatrial node, controls the heart rate and coordinates contractions to maintain a steady rhythm.

The Skin: Protection and Sensory Interface

The skin, the body's largest organ, integrates tissues to protect against environmental damage and regulate temperature. The outermost layer, the **epidermis**, is made of **stratified squamous epithelial tissue**, providing a waterproof and durable barrier. It prevents the entry of pathogens and minimizes water loss.

Beneath the epidermis, **connective tissue** in the dermis provides elasticity and strength. This layer contains blood vessels, lymphatic vessels, and sensory receptors. Specialized connective tissues, such as adipose tissue, insulate the body and cushion underlying structures.

Nervous tissue in the skin detects stimuli like pressure, pain, and temperature, transmitting this information to the brain. Muscles attached to hair follicles, known as arrector pili, contract in response to cold or fear, causing goosebumps.

The Lungs: Gas Exchange

The lungs rely on tissue integration to facilitate gas exchange. The alveoli, where oxygen and carbon dioxide are exchanged, are lined with **simple squamous**

epithelial tissue that allows rapid diffusion of gases. The surrounding **connective tissue** contains elastic fibers, enabling the lungs to expand and recoil during breathing.

Smooth muscle in the bronchi and bronchioles regulates airflow by adjusting the diameter of airways. **Nervous tissue** controls these contractions and monitors oxygen and carbon dioxide levels, ensuring efficient respiration.

The Kidneys: Filtration and Excretion

The kidneys filter blood to remove waste and maintain fluid balance. Their filtration units, the nephrons, consist of **epithelial tissue** specialized for absorption and secretion. Simple cuboidal epithelium in the tubules reabsorbs essential nutrients and ions while excreting waste into the urine.

The kidney's **connective tissue** framework supports the nephrons and houses blood vessels, while **smooth muscle** in the ureters propels urine to the bladder. **Nervous tissue** regulates kidney function, controlling blood flow and responding to signals like changes in blood pressure.

The Brain: Integration and Control

The brain, the command center of the body, relies primarily on **nervous tissue** to process and transmit information. Neurons in the brain's gray matter handle decision-making and sensory integration, while those in the white matter transmit signals to other parts of the body.

Connective tissue, in the form of meninges, protects the brain and anchors it within the skull. Blood vessels within the connective tissue deliver oxygen and nutrients. **Epithelial tissue** in the choroid plexus produces cerebrospinal fluid, which cushions the brain and removes waste.

Tissue Repair and Regeneration: The Role of Stem Cells

Tissue repair and regeneration are essential processes that restore the structure and function of damaged tissues. Stem cells have a central role in these processes due to their unique ability to divide, differentiate into specialized cell types, and self-renew. Understanding how tissues repair and regenerate offers insight into how the body maintains its integrity and heals itself after injury.

Phases of Tissue Repair

Tissue repair occurs in three overlapping phases: inflammation, proliferation, and remodeling. These phases involve a coordinated effort between various cell types, signaling molecules, and the extracellular matrix.

1. **Inflammation:** This phase begins immediately after injury, as damaged cells release chemical signals that attract immune cells to the site. Neutrophils and macrophages clear debris and pathogens through phagocytosis, preparing the tissue for repair. Inflammatory signals, such as cytokines and growth factors, activate local and recruited stem cells, priming them for regeneration.
2. **Proliferation:** Fibroblasts and stem cells dominate this phase. Fibroblasts produce collagen and other extracellular matrix components to form granulation tissue, a temporary scaffold for new cells. Stem cells differentiate into the specific cell types needed to replace lost or damaged tissue. For example, in the skin, epithelial stem cells proliferate and migrate to cover wounds, while endothelial cells form new blood vessels in a process called angiogenesis.
3. **Remodeling:** In this final phase, granulation tissue is replaced by mature tissue. Collagen fibers are reorganized, cross-linked, and aligned to restore strength and functionality. Stem cells contribute to this phase by replenishing specialized cells and maintaining tissue homeostasis.

Stem Cells: Nature's Repair System

Stem cells are undifferentiated cells with the ability to divide and produce progeny that either remain as stem cells or differentiate into specific cell types. They are classified into three main types based on their source and potency: embryonic stem cells (ESCs), adult stem cells, and induced pluripotent stem cells (iPSCs).

- **Embryonic Stem Cells (ESCs):** Derived from the inner cell mass of the blastocyst, ESCs are pluripotent, meaning they can differentiate into all cell types of the body. While ESCs have immense regenerative potential, ethical concerns and the risk of uncontrolled growth limit their clinical use.
- **Adult Stem Cells:** Also called somatic stem cells, these are found in specific tissues and are multipotent, capable of producing a limited range of cell types. For instance, hematopoietic stem cells in bone marrow generate blood cells, while mesenchymal stem cells (MSCs) differentiate into bone, cartilage, and fat cells. Adult stem cells are integral to routine tissue maintenance and repair.
- **Induced Pluripotent Stem Cells (iPSCs):** Created by reprogramming mature cells into a pluripotent state, iPSCs offer the advantages of ESCs without ethical concerns. They can be generated from a patient's own cells, reducing the risk of immune rejection in regenerative therapies.

Stem Cells in Tissue-Specific Repair

Different tissues rely on their own resident stem cell populations for repair and regeneration. These specialized stem cells ensure that tissues maintain their structure and function after injury.

- **Epithelial Tissue:** Epithelial stem cells in the skin reside in the basal layer of the epidermis and in hair follicles. After a cut or abrasion, these stem cells rapidly divide and migrate to the wound site, differentiating into keratinocytes to restore the epidermis. In the intestines, crypt base columnar cells serve as stem cells, continuously replenishing the epithelial lining that is shed daily due to mechanical and chemical stress.
- **Connective Tissue:** Mesenchymal stem cells (MSCs) in bone marrow, fat, and other tissues are essential for repairing connective tissues. MSCs can differentiate into osteoblasts to heal fractured bones, chondrocytes to repair cartilage, and adipocytes to restore fat tissue. Fibroblasts, although not true stem cells, exhibit plasticity and contribute to the repair of connective tissues by synthesizing extracellular matrix components.
- **Muscle Tissue:** Skeletal muscle regeneration depends on satellite cells, a type of stem cell located between the muscle fiber membrane and the surrounding connective tissue. Satellite cells are quiescent under normal conditions but activate after injury to proliferate and fuse with existing muscle fibers or form new ones. In cardiac muscle, however, regeneration is limited due to the low abundance of resident cardiac stem cells, resulting in scar formation after significant damage, such as a myocardial infarction.
- **Nervous Tissue:** Neural stem cells (NSCs) are found in specific regions of the brain, such as the subventricular zone and the hippocampus. These cells generate neurons, astrocytes, and oligodendrocytes, contributing to neurogenesis in response to injury. However, the nervous system has limited regenerative capacity, especially in the spinal cord. Research is exploring ways to harness NSCs or transplant iPSCs to repair damaged neural tissue.

Stem Cell Niche and Microenvironment

The **stem cell niche** is the microenvironment where stem cells reside. This niche provides physical support and biochemical signals that regulate stem cell behavior, such as self-renewal, differentiation, and migration. Components of the niche include extracellular matrix proteins, neighboring cells, and soluble factors like growth factors and cytokines.

For example, in the bone marrow, hematopoietic stem cells interact with stromal cells, endothelial cells, and osteoblasts, which secrete signals that maintain stem cell quiescence or activate them during injury. Disruption of the niche can impair tissue repair, emphasizing the importance of the microenvironment in stem cell function.

Challenges in Regeneration

Tissue repair and regeneration are not always efficient, and some tissues, such as cartilage and nervous tissue, have limited regenerative capacity. Factors that affect regeneration include the severity of the injury, the age of the individual, and the presence of chronic diseases like diabetes. In such cases, fibrosis, or scar formation, replaces functional tissue, compromising organ performance.

Excessive fibrosis is a significant challenge. For instance, in the liver, chronic damage from alcohol abuse or viral hepatitis leads to cirrhosis, where fibrotic tissue replaces hepatocytes. Understanding the balance between regeneration and fibrosis is critical for developing effective therapies.

Therapeutic Applications of Stem Cells

The regenerative potential of stem cells has inspired the development of therapies for a wide range of conditions. Stem cell transplants are already used in treating blood disorders like leukemia, where hematopoietic stem cells from bone marrow or cord blood replace diseased cells.

For cartilage repair, MSCs are being explored to regenerate damaged joint tissue, potentially offering a solution for osteoarthritis. In cardiac therapy, researchers are investigating ways to use stem cells to restore heart tissue after a heart attack, aiming to improve function and reduce scarring.

In the nervous system, stem cell therapies are being tested for conditions such as spinal cord injuries, Parkinson's disease, and multiple sclerosis. Transplanting neural stem cells or iPSCs may help replace lost neurons and support repair processes.

Tissue Engineering and Stem Cells

Tissue engineering combines stem cells with biomaterials to create scaffolds that mimic the extracellular matrix, providing a framework for cell growth and tissue formation. These scaffolds can be implanted to repair tissues or used in labs to grow organs for transplantation. For example, engineered skin grafts using epithelial stem cells are already being used to treat severe burns.

Advances in 3D bioprinting allow precise placement of stem cells and biomaterials to construct tissues layer by layer. Researchers have successfully printed cartilage, bone, and liver tissue, moving closer to creating fully functional organs for transplantation.

Future Directions in Regenerative Medicine

The integration of stem cell research with technologies like CRISPR enables the correction of genetic mutations and the generation of patient-specific cells for transplantation, reducing rejection risks. Organoid technology advances tissue modeling and drug testing, showcasing stem cells' transformative role in understanding and repairing human tissues.

CHAPTER 6: THE SKELETAL SYSTEM

Structure and Function of Bones

Bones form the framework of the human body, providing structure, protection, and support for movement. They are living tissues, capable of growth, remodeling, and repair. Bones interact with muscles, ligaments, and tendons to enable movement, and they also serve critical roles in mineral storage and blood cell production. Their complex structure allows them to be both lightweight and strong, accommodating the demands of everyday activities and physical stress.

Bone Composition and Structure

Bones are composed of both organic and inorganic materials. The organic matrix, primarily collagen, gives bones flexibility and tensile strength, allowing them to resist breaking under tension. The inorganic portion, made up of hydroxyapatite (a crystalline form of calcium phosphate), provides rigidity and compressive strength. This combination of collagen and minerals makes bones resilient and durable.

The structure of bone can be divided into two types: **compact bone** and **spongy bone**. Compact bone forms the dense, outer layer, while spongy bone, also called cancellous bone, is found inside, particularly at the ends of long bones and in the interior of flat bones like the sternum. Spongy bone contains a lattice-like network of trabeculae, which reduces weight while maintaining strength.

Gross Anatomy of Bones

Bones come in various shapes and sizes, classified as long, short, flat, irregular, or sesamoid.

- **Long bones**, such as the femur and humerus, are longer than they are wide. They consist of a shaft (diaphysis) and expanded ends (epiphyses). The diaphysis is hollow, housing the medullary cavity filled with yellow marrow, while the epiphyses are covered in spongy bone and red marrow.
- **Short bones**, like those in the wrist and ankle, are roughly cube-shaped, providing stability with limited movement.
- **Flat bones**, such as the skull, ribs, and scapulae, protect internal organs and offer surfaces for muscle attachment.
- **Irregular bones**, like the vertebrae and pelvis, have complex shapes suited to their functions.
- **Sesamoid bones**, such as the patella, develop within tendons and reduce friction during joint movement.

The outer surface of bones is covered by the **periosteum**, a dense connective tissue layer rich in blood vessels and nerves. The periosteum contains osteoblasts, which are essential for bone growth and repair. Inside, the bone is lined with the **endosteum**, a thin layer of cells that also contribute to bone remodeling.

Microscopic Structure

The microscopic structure of compact bone is organized into units called **osteons**, or Haversian systems. Each osteon consists of concentric lamellae, or layers of bone matrix, surrounding a central canal. The central canal houses blood vessels and nerves that supply the bone. Small channels called **canaliculi** connect osteocytes (mature bone cells) within lacunae, allowing the exchange of nutrients and waste.

Spongy bone lacks osteons but has trabeculae, which are oriented along lines of stress. This arrangement efficiently distributes mechanical loads and protects the marrow contained within.

Bone Cells and Their Roles

Bone tissue is dynamic, maintained by the coordinated activity of several cell types:

- **Osteoblasts** are responsible for forming new bone. They secrete collagen and other components of the bone matrix, which later become mineralized.
- **Osteocytes** are mature osteoblasts embedded within the bone matrix. They maintain bone tissue and communicate mechanical stress signals to osteoblasts and osteoclasts, ensuring appropriate remodeling.
- **Osteoclasts** break down bone tissue through resorption. These large, multinucleated cells release enzymes and acids that dissolve the mineral and organic components of bone, a process essential for calcium homeostasis and bone remodeling.

Functions of Bones

1. **Support and Protection:** Bones provide a rigid framework that supports the body and protects vital organs. The skull encloses the brain, the rib cage shields the heart and lungs, and the vertebrae protect the spinal cord. These protective structures ensure that critical systems are shielded from mechanical damage.

2. **Movement:** Bones act as levers, and joints serve as fulcrums. When muscles contract, they pull on bones, producing movement. For example, the femur and tibia form a hinge joint at the knee, allowing bending and straightening motions. The arrangement of bones and muscles determines the range and strength of movements.

3. **Mineral Storage:** Bones store essential minerals, particularly calcium and phosphate, which are vital for physiological processes like nerve conduction, muscle contraction, and blood clotting. When blood calcium levels drop, osteoclasts resorb bone to release calcium into the bloodstream. Conversely, when calcium levels are high, osteoblasts deposit calcium into the bone matrix.

4. **Blood Cell Production:** The red marrow within spongy bone is the site of hematopoiesis, the process of forming blood cells. Red blood cells, white blood cells, and platelets originate from hematopoietic stem cells in the marrow. This function is most active in the flat bones, such as the sternum and pelvis, and the ends of long bones in children and adults.

5. **Energy Storage:** Yellow marrow in the medullary cavity stores lipids, providing an energy reserve. While primarily inactive compared to red marrow, yellow marrow can revert to red marrow under conditions of severe blood loss or increased demand for blood cell production.

Bone Growth and Development

Bone development begins during embryogenesis through two processes: **intramembranous ossification** and **endochondral ossification**.

- In intramembranous ossification, flat bones like the skull and clavicles form directly from mesenchymal tissue. Osteoblasts differentiate from mesenchymal cells, deposit bone matrix, and create spongy bone that later becomes compact bone.
- In endochondral ossification, most bones form from a cartilage template. Chondrocytes (cartilage cells) proliferate, hypertrophy, and are replaced by bone-forming osteoblasts. This process occurs at growth plates in long bones, allowing for elongation during childhood and adolescence.

Growth plates, or epiphyseal plates, remain active until the end of puberty, after which they ossify, and longitudinal growth ceases.

Bone Remodeling

Bone remodeling is a continuous process in which old or damaged bone is replaced with new bone. This process is essential for maintaining bone strength, adapting to mechanical stress, and regulating calcium levels.

The remodeling cycle involves the coordination of osteoclasts and osteoblasts. Osteoclasts resorb bone, creating small cavities, while osteoblasts fill these cavities with new matrix. Weight-bearing activities stimulate remodeling, strengthening areas of bone that experience the most stress. Conversely, lack of physical activity leads to bone loss, a phenomenon observed in conditions like osteoporosis.

Factors Affecting Bone Health

Bone health is influenced by genetics, nutrition, hormonal balance, and physical activity. Adequate intake of **calcium** and **vitamin D** is crucial for bone mineralization. Calcium provides the raw material for hydroxyapatite, while vitamin D enhances calcium absorption in the intestines. Deficiencies in these nutrients can lead to conditions like rickets in children or osteomalacia in adults, characterized by weak and deformed bones.

Hormones like **parathyroid hormone (PTH)**, **calcitonin**, and **estrogen** regulate bone remodeling. PTH increases bone resorption to raise blood calcium levels, while calcitonin promotes bone formation by inhibiting osteoclast activity. Estrogen helps maintain bone density by suppressing osteoclasts, explaining why postmenopausal women are at higher risk of osteoporosis due to declining estrogen levels.

Regular weight-bearing exercise, such as walking or resistance training, promotes bone formation and prevents bone loss. Mechanical stress stimulates osteocytes to signal for increased bone remodeling, resulting in stronger bones.

Bone's Role in the Endocrine System

Bones are not merely structural elements; they also function as endocrine organs. Osteoblasts produce a hormone called **osteocalcin**, which influences energy metabolism by regulating insulin secretion and sensitivity. This discovery has broadened our understanding of the interconnectedness of the skeletal system with other physiological systems.

The structure and function of bones illustrate their versatility and adaptability. Bones are not static; they respond dynamically to the body's needs, providing a foundation for movement, protection, and overall health.

Bone Development and Growth: Ossification

Bone development and growth are highly regulated processes that begin in the embryo and continue through adolescence, ensuring that the skeletal system supports the body's structure and function. The formation of bone, known as **ossification**, occurs in two primary ways: intramembranous ossification and endochondral ossification. Both processes involve the transformation of precursor tissues into bone, but they differ in their mechanisms and locations.

Intramembranous Ossification

Intramembranous ossification forms flat bones, such as the skull, mandible, and clavicles. This process begins in the embryonic mesenchyme, a type of connective tissue composed of undifferentiated cells.

Mesenchymal cells cluster and differentiate into **osteoblasts**, which secrete bone matrix (osteoid). The osteoid calcifies, trapping the osteoblasts within it and transforming them into **osteocytes**, the mature bone cells that maintain the bone matrix. The calcified regions form spicules that grow outward, connecting to form a network of **trabeculae**. Blood vessels infiltrate this developing spongy bone, supporting the deposition of additional bone material.

As the process continues, the outer layers of spongy bone are compacted, forming the protective outer shell of compact bone. The internal regions remain spongy, providing a lightweight structure that can withstand stress while housing bone marrow.

Endochondral Ossification

Endochondral ossification is the predominant method of bone formation, responsible for the development of long bones, such as the femur, tibia, and humerus. It begins with a **cartilage model** made of hyaline cartilage, which provides a template for bone formation.

During early fetal development, chondrocytes (cartilage cells) proliferate and enlarge within the cartilage model. These hypertrophic chondrocytes begin to secrete enzymes that calcify the surrounding cartilage matrix, leading to their death as nutrients can no longer diffuse through the hardened matrix. This creates a scaffold for incoming bone cells.

The process advances when a **periosteal bone collar** forms around the diaphysis (shaft) of the cartilage model. Blood vessels invade the center of the shaft, bringing osteoblasts and osteoclasts. These cells replace the calcified cartilage with bone, creating the **primary ossification center** in the diaphysis.

As the bone grows, **secondary ossification centers** develop in the epiphyses (ends of the bone). Unlike the diaphysis, some cartilage is retained in the epiphyses as **articular cartilage**, which cushions joints, and as the **epiphyseal plate**, also known as the growth plate.

Bone Growth and Lengthening

Bone lengthening occurs at the **epiphyseal plate**, where cartilage cells divide and expand, pushing the epiphysis away from the diaphysis. This region has distinct zones:

- **Resting zone:** Anchors the epiphyseal plate to the bone.
- **Proliferative zone:** Chondrocytes divide rapidly, forming stacks of cells.

- **Hypertrophic zone:** Chondrocytes enlarge, accumulating nutrients and signaling calcification.
- **Calcification zone:** The cartilage matrix calcifies, and chondrocytes die.
- **Ossification zone:** Osteoblasts replace the calcified cartilage with bone tissue.

This process continues until the end of puberty, when hormonal changes cause the epiphyseal plates to ossify, leaving an **epiphyseal line** and halting further growth in length.

Bone Growth in Width

Bones also grow in width through **appositional growth**, which occurs at the surface of the bone. Osteoblasts in the periosteum deposit new bone matrix on the external surface, while osteoclasts in the endosteum resorb bone from the inner surface. This balanced activity maintains the bone's shape while increasing its diameter.

Factors Influencing Bone Growth

Bone growth is influenced by **genetics, nutrition, hormones, and physical activity**. Adequate intake of calcium and vitamin D is essential for mineralization, while proteins and vitamins like C and K support collagen synthesis. Hormones such as **growth hormone, thyroxine,** and **sex hormones** (estrogen and testosterone) regulate growth plate activity and bone density.

Mechanical stress from physical activity stimulates bone remodeling, enhancing strength. Conversely, inactivity can lead to bone thinning, increasing the risk of fractures.

The Role of Joints and Cartilage

Joints and cartilage are essential components of the skeletal system, enabling movement, providing flexibility, and reducing friction between bones. While bones form the rigid framework, joints and cartilage allow for mobility and adaptability, ensuring the body functions efficiently during various physical activities.

Types of Joints

Joints, also called articulations, are classified based on their structure and the amount of movement they allow. There are three main types:

1. **Fibrous Joints:** These joints are connected by dense connective tissue and allow little to no movement.

- - **Sutures** in the skull fuse bones tightly, protecting the brain.
  - **Syndesmoses**, such as the connection between the tibia and fibula, provide slight movement and stability.
  - **Gomphoses**, found in the teeth sockets, secure teeth in place.

2. **Cartilaginous Joints:** These joints are connected by cartilage and allow limited movement.

   - **Synchondroses** involve hyaline cartilage, such as the epiphyseal plates during growth.
   - **Symphyses**, like the pubic symphysis or intervertebral discs, consist of fibrocartilage, providing strength and shock absorption.

3. **Synovial Joints:** These highly movable joints are the most common and versatile in the body. They include:

   - **Hinge joints** (e.g., elbow, knee) that allow flexion and extension.
   - **Ball-and-socket joints** (e.g., shoulder, hip) that enable rotation and multi-directional movement.
   - **Pivot joints** (e.g., atlantoaxial joint in the neck) that permit rotational movement.
   - **Saddle joints** (e.g., thumb joint) that allow angular movement with a greater range than hinge joints.
   - **Plane joints** (e.g., between carpals) that enable sliding movements.
   - **Condyloid joints** (e.g., wrist joint) that allow movement in two planes.

Synovial joints are characterized by a **synovial cavity** filled with fluid, which lubricates and nourishes the cartilage, minimizing friction during movement.

The Function of Cartilage in Joints

Cartilage is a flexible, resilient tissue that supports joint function and absorbs impact. It comes in three types, each adapted for specific roles:

1. **Hyaline Cartilage:** Found in articular surfaces of synovial joints, it provides a smooth, low-friction surface that cushions bones during movement. Hyaline cartilage is also present in the rib cage, nose, and trachea.

2. **Elastic Cartilage:** Found in structures like the ear and epiglottis, it is more flexible than hyaline cartilage, maintaining shape while allowing deformation.

3. **Fibrocartilage:** Found in intervertebral discs, the pubic symphysis, and menisci of the knee, fibrocartilage is tough and resists compression and

shear forces. It acts as a shock absorber, protecting joints during high-impact activities.

Cartilage has a limited capacity for repair due to its avascular nature. Nutrients and waste products are exchanged via diffusion, a slower process compared to vascularized tissues. This limitation makes cartilage damage from injury or wear-and-tear difficult to heal, often leading to conditions like osteoarthritis.

Ligaments and Tendons

Ligaments and tendons are connective tissues associated with joints, providing stability and facilitating movement. **Ligaments** connect bones to other bones, reinforcing joints and preventing excessive movement. For example, the anterior cruciate ligament (ACL) stabilizes the knee by limiting forward motion of the tibia relative to the femur.

Tendons connect muscles to bones, transmitting the force generated by muscle contractions to create movement. The Achilles tendon, for instance, allows the calf muscles to move the foot for walking or running.

Joint Disorders and Cartilage Degeneration

Joint and cartilage health are vital for maintaining mobility. Disorders such as **osteoarthritis** result from the gradual breakdown of articular cartilage, leading to pain, stiffness, and decreased range of motion. Rheumatoid arthritis, an autoimmune disease, causes inflammation of the synovial membrane, damaging both cartilage and bone over time.

Injuries such as meniscal tears or ligament sprains can destabilize joints, requiring surgical repair or prolonged rehabilitation. Advances in tissue engineering and stem cell therapy offer hope for regenerating damaged cartilage and restoring joint function.

Calcium Homeostasis and Bone Health

Calcium homeostasis is essential for the proper functioning of the skeletal system and the body as a whole. Bones serve as the primary reservoir for calcium, housing about 99% of the body's total calcium stores. This mineral is critical for maintaining bone strength and supporting various physiological processes, including muscle contraction, nerve transmission, blood clotting, and enzyme activity. The skeletal system dynamically regulates calcium levels through continuous remodeling, ensuring both bone health and stable calcium concentrations in the bloodstream.

The Role of Calcium in Bone

Calcium provides the structural integrity of bones by forming hydroxyapatite crystals, which combine with collagen in the bone matrix to create a rigid and durable structure. This mineralization process allows bones to withstand compression and support the body's weight. Inadequate calcium availability disrupts this balance, leading to weakened bones and increased susceptibility to fractures.

Calcium is not permanently fixed in the bone; it is continuously exchanged between bone tissue and the bloodstream. This exchange is a critical component of **bone remodeling**, which involves osteoblasts building new bone and osteoclasts breaking down old bone. These cells work together to adapt to mechanical stress, repair microdamage, and regulate calcium availability.

Hormonal Regulation of Calcium Homeostasis

The balance of calcium in the body is tightly regulated by hormones, primarily **parathyroid hormone (PTH), calcitonin,** and **vitamin D**. These hormones maintain blood calcium levels within a narrow range, typically between 8.5 and 10.5 mg/dL.

- **Parathyroid Hormone (PTH):** Secreted by the parathyroid glands when blood calcium levels drop, PTH increases calcium levels by stimulating osteoclast activity, which releases calcium from bone into the bloodstream. PTH also enhances calcium reabsorption in the kidneys and promotes the activation of vitamin D, further boosting calcium absorption in the intestines.

- **Calcitonin:** Produced by the thyroid gland, calcitonin counteracts the effects of PTH by inhibiting osteoclast activity and encouraging calcium deposition in bone. This hormone is more active during periods of high calcium levels, helping to protect the skeleton from excessive resorption.

- **Vitamin D:** Activated vitamin D (calcitriol) enhances calcium absorption in the intestines and facilitates its incorporation into bone. Without sufficient vitamin D, calcium absorption is inefficient, leading to low blood calcium levels despite adequate dietary intake.

Dietary Calcium and Bone Health

Calcium intake is vital for maintaining bone health throughout life. The recommended daily intake of calcium varies by age and physiological state, with adolescents, pregnant individuals, and older adults requiring higher amounts. Dietary sources of calcium include dairy products, leafy green vegetables, fortified foods, and certain fish like salmon and sardines. For individuals with low calcium intake, supplements may be necessary to meet daily requirements.

In addition to calcium, **vitamin D** is essential for calcium metabolism. Sunlight exposure triggers the production of vitamin D in the skin, but dietary sources such as fortified milk, fatty fish, and egg yolks are also important. Without adequate vitamin D, even a calcium-rich diet cannot effectively maintain bone density.

Bone Resorption and Calcium Mobilization

When blood calcium levels are low, bone resorption increases to release calcium into the bloodstream. This process is regulated by osteoclasts, which dissolve the mineralized matrix and degrade collagen. While resorption is necessary for calcium homeostasis, excessive or prolonged bone loss can weaken the skeleton.

Conditions like **osteoporosis** occur when bone resorption outpaces bone formation, leading to reduced bone mass and structural deterioration. This imbalance is influenced by aging, hormonal changes, and dietary deficiencies. Postmenopausal women are particularly susceptible due to decreased estrogen levels, which normally inhibit osteoclast activity.

Calcium Storage and Buffering

Bone acts as a reservoir to buffer fluctuations in blood calcium levels. When calcium intake is sufficient, bones store excess calcium, maintaining skeletal strength. During periods of calcium deficiency, this stored calcium is mobilized to meet the body's needs. However, chronic calcium depletion leads to progressive bone loss and increased fracture risk.

This buffering system highlights the dual role of bones as both structural support and a dynamic mineral reservoir. The body prioritizes calcium homeostasis over bone strength, ensuring adequate calcium for critical physiological functions even at the expense of bone density.

Physical Activity and Calcium Utilization

Physical activity stimulates bone remodeling and improves calcium retention in bone tissue. Weight-bearing exercises, such as walking, running, and resistance training, generate mechanical stress on the skeleton, prompting osteoblasts to deposit new bone. These activities increase bone density and reduce the risk of fractures, particularly in weight-bearing bones like the femur and vertebrae.

In contrast, sedentary lifestyles or extended periods of immobilization lead to reduced bone formation and accelerated resorption. This phenomenon is evident in astronauts who experience bone loss during prolonged weightlessness, as well as in bedridden individuals.

Calcium-Related Disorders

Disruptions in calcium homeostasis can result in a variety of skeletal and systemic disorders:

- **Hypocalcemia:** Low blood calcium levels impair bone mineralization and can lead to conditions like **rickets** in children and **osteomalacia** in adults. These disorders are characterized by soft, weak bones prone to deformities. Hypocalcemia also causes muscle spasms, tingling, and seizures due to increased neuromuscular excitability.

- **Hypercalcemia:** Excessive blood calcium levels may result from overactive parathyroid glands (hyperparathyroidism), excessive vitamin D intake, or malignancies. Hypercalcemia reduces bone density and causes symptoms like kidney stones, fatigue, and irregular heart rhythms.

- **Osteoporosis:** A condition where decreased bone density increases fracture risk. While not directly caused by calcium deficiency, inadequate calcium and vitamin D intake contribute to its progression. Osteoporosis commonly affects older adults, particularly postmenopausal women and individuals with sedentary lifestyles.

Strategies for Maintaining Calcium Homeostasis

Maintaining optimal calcium levels requires a combination of proper nutrition, physical activity, and hormonal regulation. Ensuring adequate dietary intake of calcium and vitamin D is fundamental, while regular exercise helps strengthen bones and reduce calcium loss. Hormonal therapies, such as estrogen replacement or bisphosphonates, are used to treat osteoporosis by reducing bone resorption.

Emerging therapies focus on regulating bone remodeling. For instance, drugs like **denosumab**, a monoclonal antibody, inhibit osteoclast activity, preserving bone density. Other approaches aim to stimulate osteoblasts, enhancing bone formation and improving overall skeletal health.

CHAPTER 7: THE MUSCULAR SYSTEM

Types of Muscles: Skeletal, Cardiac, and Smooth

Muscles are essential for movement, stability, and the functioning of internal organs. They convert chemical energy from ATP into mechanical force, allowing the body to move, pump blood, and maintain internal processes. There are three types of muscles in the human body: skeletal, cardiac, and smooth. Each type has unique structures, locations, and functions that suit their specific roles.

Skeletal Muscle: Voluntary Movement and Strength

Skeletal muscle is attached to bones via tendons, enabling voluntary movements such as walking, running, and lifting objects. These muscles are also involved in maintaining posture, stabilizing joints, and generating heat during physical activity. Skeletal muscles are under conscious control, meaning their contractions are directed by the nervous system.

Structurally, skeletal muscle fibers are long, cylindrical, and multinucleated. The fibers are striated, with alternating light and dark bands visible under a microscope. These striations arise from the highly organized arrangement of **sarcomeres**, the functional units of muscle contraction. Sarcomeres are composed of overlapping **actin** (thin filaments) and **myosin** (thick filaments), which slide past each other during contraction.

The contraction of skeletal muscles is initiated by signals from motor neurons. These neurons release acetylcholine at the neuromuscular junction, triggering an electrical impulse in the muscle fiber. The impulse travels along the sarcolemma (the cell membrane of the muscle fiber) and into the **T-tubules**, leading to the release of calcium ions from the sarcoplasmic reticulum. Calcium binds to **troponin**, causing a conformational change that moves tropomyosin away from actin's binding sites, allowing myosin heads to attach and generate a contraction through the power stroke.

Skeletal muscles can be classified into two main types based on their contraction speed and endurance:

1. **Slow-twitch fibers (Type I):** These fibers are fatigue-resistant and specialize in sustained, aerobic activities like long-distance running. They have abundant mitochondria and rely on oxidative phosphorylation for energy.

2. **Fast-twitch fibers (Type II):** These fibers contract quickly and generate more force but fatigue rapidly. They are suited for short bursts of power, such as sprinting or weightlifting.

The body's ability to adapt to exercise, whether through hypertrophy (muscle growth) or increased endurance, demonstrates the plasticity of skeletal muscle. Strength training increases the size and force of individual fibers, while aerobic exercise enhances their oxidative capacity.

Cardiac Muscle: The Heart's Pump

Cardiac muscle is found exclusively in the walls of the heart, where it contracts rhythmically to pump blood throughout the body. Unlike skeletal muscle, cardiac muscle operates involuntarily, controlled by the autonomic nervous system and intrinsic pacemaker cells.

Cardiac muscle fibers are striated, similar to skeletal muscle, but they are shorter, branched, and typically uninucleated. These cells are connected by **intercalated discs**, specialized structures that contain gap junctions and desmosomes. Gap junctions allow the rapid spread of electrical impulses between cells, ensuring that the heart contracts as a coordinated unit. Desmosomes provide mechanical strength, preventing the cells from pulling apart during the forceful contractions of the heartbeat.

The rhythmic contractions of cardiac muscle are regulated by the **sinoatrial (SA) node**, the heart's natural pacemaker. The SA node generates electrical impulses that propagate through the heart's conduction system, triggering contractions in a precise sequence: first the atria, then the ventricles. This coordinated activity ensures efficient blood flow and oxygen delivery to the body.

Cardiac muscle relies almost entirely on aerobic metabolism, which makes it highly resistant to fatigue. It contains numerous mitochondria and a rich blood supply to meet its continuous energy demands. However, prolonged oxygen deprivation, as seen in a heart attack, can damage cardiac muscle cells, leading to impaired function.

Smooth Muscle: Involuntary Control of Internal Organs

Smooth muscle is found in the walls of hollow organs, including blood vessels, the digestive tract, the respiratory passages, and the bladder. It controls involuntary movements, such as the constriction and dilation of blood vessels, the movement of food through the intestines, and the regulation of airflow in the bronchi.

Unlike skeletal and cardiac muscle, smooth muscle lacks striations because its actin and myosin filaments are not arranged in sarcomeres. Instead, the filaments are organized in a crisscross pattern throughout the cell, anchored to **dense bodies** in the cytoplasm. When the muscle contracts, the dense bodies pull the cell into a spindle shape.

Smooth muscle cells are small, spindle-shaped, and uninucleated. They operate under involuntary control, receiving signals from the autonomic nervous system, hormones, or local factors like changes in pH or oxygen levels. For example, smooth muscle in the walls of arterioles constricts in response to sympathetic stimulation, increasing blood pressure.

Smooth muscle contractions are slower and more sustained than those of skeletal or cardiac muscle. This efficiency is critical for maintaining functions like blood vessel tone and peristalsis (the wave-like movement that propels food through the digestive tract). Smooth muscle can also remain contracted for extended periods without significant energy expenditure, a property known as **latch-state** mechanics.

Smooth muscle is further divided into two types based on how the cells communicate:

1. **Single-unit smooth muscle:** Found in the walls of the intestines and blood vessels, single-unit smooth muscle cells are connected by gap junctions, allowing them to contract as a coordinated sheet.
2. **Multi-unit smooth muscle:** Found in the iris of the eye and the walls of large airways, multi-unit smooth muscle cells contract independently, enabling fine control.

Comparison of Muscle Types

Skeletal, cardiac, and smooth muscles share the ability to contract, but their structures, control mechanisms, and functions differ significantly.

- **Skeletal muscle** provides voluntary control, enabling precise and forceful movements. It is striated, multinucleated, and relies on nervous system input for contraction.
- **Cardiac muscle** operates autonomously, with rhythmic, coordinated contractions driven by intrinsic pacemaker cells. It is striated but branched, and its intercalated discs ensure synchronization.
- **Smooth muscle** functions involuntarily, sustaining contractions over long periods to regulate internal processes. It is non-striated, spindle-shaped, and highly efficient.

Each muscle type is adapted to its specific role, contributing to the overall functioning of the body. For instance, skeletal muscle moves the body in response to conscious decisions, cardiac muscle keeps the heart beating reliably, and smooth muscle maintains the flow of blood, air, and digestive contents. Together, these muscle types create a system capable of both powerful, deliberate actions and subtle, continuous regulation.

Muscle Contraction: The Role of Actin, Myosin, and Calcium

Muscle contraction is a highly coordinated process that transforms chemical energy into mechanical force, enabling movement and stability. This process depends on the interaction between **actin** and **myosin**, two proteins that form the contractile machinery of muscle fibers, and the regulation of these interactions by **calcium ions**. Understanding the molecular mechanisms of muscle contraction provides insight into how the muscular system generates force and adapts to different demands.

The Structure of Sarcomeres

The sarcomere is the functional unit of muscle contraction. It is composed of overlapping filaments of actin (thin filaments) and myosin (thick filaments), arranged in a precise, repeating pattern. Each sarcomere is bordered by **Z-discs**, which anchor the actin filaments. Myosin filaments are centrally located and partially overlap with actin filaments, creating the striated appearance of skeletal and cardiac muscle.

The regions of the sarcomere change during contraction. The **A-band**, which corresponds to the length of the myosin filaments, remains constant, while the **I-band** (actin-only region) and **H-zone** (myosin-only region) shrink as the actin filaments slide past the myosin filaments, shortening the sarcomere and generating tension.

The Sliding Filament Theory

The sliding filament theory explains how actin and myosin interact to produce contraction. Myosin filaments have protruding **myosin heads**, which act as molecular motors. These heads attach to actin filaments and pull them toward the center of the sarcomere through a process called the **power stroke**.

1. **Cross-bridge Formation:** In a resting muscle, myosin heads are energized by the hydrolysis of ATP into ADP and inorganic phosphate (Pi). When calcium binds to the regulatory protein **troponin** on the actin filament, it induces a conformational change that moves **tropomyosin**, exposing binding sites on actin. Myosin heads attach to these sites, forming cross-bridges.

2. **Power Stroke:** The release of ADP and Pi from the myosin head triggers the power stroke. The myosin head pivots, pulling the actin filament toward the center of the sarcomere. This movement generates the force of contraction.

3. **Detachment:** A new ATP molecule binds to the myosin head, causing it to detach from actin. Without ATP, as seen in rigor mortis, myosin remains bound to actin, resulting in stiffness.

4. **Reactivation:** ATP is hydrolyzed into ADP and Pi, re-energizing the myosin head and positioning it for the next cycle.

The Role of Calcium in Contraction

Calcium ions act as the on/off switch for muscle contraction. In resting muscle, calcium levels in the cytoplasm are low, and tropomyosin blocks the binding sites on actin. Upon stimulation, an action potential travels along the sarcolemma and into the **T-tubules**, triggering the release of calcium from the **sarcoplasmic reticulum (SR)** into the cytoplasm.

Calcium binds to **troponin-C**, causing the troponin complex to shift tropomyosin away from actin's binding sites. This exposure allows myosin to interact with actin, initiating contraction. As the signal ends, calcium is actively pumped back into the SR by **calcium ATPase pumps**, reducing cytoplasmic calcium levels and allowing tropomyosin to re-cover actin, stopping the contraction.

Excitation-Contraction Coupling

Excitation-contraction coupling describes the sequence of events that links the arrival of a neural signal to muscle contraction. It begins with an action potential in a motor neuron, which releases **acetylcholine (ACh)** at the neuromuscular junction. ACh binds to receptors on the muscle fiber, triggering an action potential in the sarcolemma. This electrical signal propagates into the T-tubules, causing the SR to release calcium, ultimately leading to contraction.

This process ensures that muscle contraction is tightly controlled and occurs only when stimulated by the nervous system.

Exercise and Muscle Health

Exercise profoundly impacts muscle health, promoting strength, endurance, and metabolic efficiency. Regular physical activity enhances the structure and function of muscle tissue while reducing the risk of chronic diseases. Different types of exercise—resistance training and aerobic exercise—have distinct effects on muscle adaptation and overall health.

Resistance Training and Muscle Strength

Resistance training, such as weightlifting, improves muscle strength and size through a process called **hypertrophy**. During resistance exercise, muscles experience microscopic tears, particularly in their sarcomeres. This mechanical stress activates **satellite cells**, a type of muscle stem cell, which proliferate and fuse with existing muscle fibers to repair damage and add new myonuclei. The increased number of myonuclei allows the muscle to produce more proteins, resulting in larger and stronger fibers.

Resistance training primarily affects **fast-twitch muscle fibers (Type II)**, which are responsible for generating power and speed. These fibers become thicker and more capable of producing force. Over time, this adaptation leads to improved performance in strength-based activities, such as lifting and sprinting.

Aerobic Exercise and Endurance

Aerobic exercise, such as running, cycling, or swimming, enhances muscular endurance and cardiovascular health. Unlike resistance training, aerobic exercise targets **slow-twitch fibers (Type I)**, which are optimized for sustained, low-intensity activities. These fibers develop more mitochondria and capillaries, improving their ability to generate ATP through oxidative phosphorylation.

Regular aerobic exercise increases oxygen delivery to muscles, enhancing their efficiency and delaying the onset of fatigue. It also improves the heart's pumping capacity and boosts overall blood flow, benefiting not only muscle health but the entire body.

Muscle Recovery and Adaptation

Recovery is an essential part of muscle adaptation. During rest, muscle fibers repair damage caused by exercise and grow stronger. Adequate **nutrition**, including protein intake, supports the synthesis of new muscle proteins, while carbohydrates replenish glycogen stores depleted during physical activity.

Stretching and flexibility exercises help maintain muscle elasticity and reduce the risk of injury. Techniques like foam rolling and massage can relieve tension and improve blood flow to recovering muscles.

Benefits of Exercise Beyond Muscles

Exercise benefits muscle health and supports broader physiological functions. Regular activity increases insulin sensitivity, enhancing glucose uptake by muscle cells and reducing the risk of type 2 diabetes. It also boosts metabolism, aiding in weight management and reducing fat accumulation.

Additionally, exercise helps preserve muscle mass during aging, a critical factor in preventing **sarcopenia**, the gradual loss of muscle strength and function. By maintaining strong muscles, older adults can improve mobility, balance, and independence.

Both resistance and aerobic exercise contribute to long-term muscle health, supporting the muscular system's ability to adapt, perform, and maintain overall vitality.

CHAPTER 8: THE NERVOUS SYSTEM

Anatomy of the Nervous System: Central and Peripheral Divisions

The nervous system is a highly organized network that coordinates the body's activities by transmitting signals between different parts of the body. It is divided into two main components: the **central nervous system (CNS)** and the **peripheral nervous system (PNS)**. These divisions work together to process sensory information, execute motor commands, and regulate involuntary functions, ensuring the body operates efficiently.

The Central Nervous System (CNS)

The central nervous system consists of the brain and spinal cord, which act as the command center for the body. It integrates sensory input, processes information, and generates responses.

The Brain

The brain is a complex organ composed of billions of neurons and glial cells. It is protected by the **skull**, cushioned by **cerebrospinal fluid (CSF)**, and enclosed within the meninges, a three-layered protective membrane. The brain is divided into three main regions: the cerebrum, cerebellum, and brainstem.

1. **Cerebrum:** The cerebrum is the largest part of the brain and is divided into two hemispheres connected by the **corpus callosum**, a bundle of nerve fibers that facilitates communication between them. Each hemisphere is further divided into four lobes:

    - **Frontal lobe**: Involved in decision-making, problem-solving, and voluntary motor functions.
    - **Parietal lobe**: Processes sensory information, such as touch, temperature, and pain.
    - **Temporal lobe**: Handles auditory perception and memory formation.
    - **Occipital lobe**: Responsible for visual processing.

2. The surface of the cerebrum, the **cerebral cortex**, is highly folded to increase surface area, allowing for greater cognitive capacity. The cortex contains regions dedicated to motor control, sensory processing, and higher-order functions like language and reasoning.

3. **Cerebellum:** Located beneath the cerebrum, the cerebellum coordinates balance, posture, and fine motor movements. It ensures that voluntary movements are smooth and precise by integrating sensory inputs with motor outputs.

4. **Brainstem:** The brainstem connects the brain to the spinal cord and regulates essential functions like breathing, heart rate, and blood pressure. It consists of the midbrain, pons, and medulla oblongata. The brainstem also houses nuclei for cranial nerves, which control sensory and motor functions of the head and neck.

The Spinal Cord

The spinal cord extends from the brainstem through the vertebral column, serving as a communication highway between the brain and the body. It is segmented into cervical, thoracic, lumbar, sacral, and coccygeal regions, each giving rise to spinal nerves.

The spinal cord contains **gray matter** in its center, shaped like a butterfly, surrounded by **white matter**. Gray matter houses neuron cell bodies and synapses, while white matter consists of myelinated axons that transmit signals. The spinal cord's main functions include:

- Relaying sensory information from the PNS to the brain.
- Transmitting motor commands from the brain to muscles and glands.
- Mediating reflexes, which are rapid, involuntary responses to stimuli.

The Peripheral Nervous System (PNS)

The peripheral nervous system connects the CNS to the rest of the body. It includes all nerves outside the brain and spinal cord, divided into the **somatic nervous system** and the **autonomic nervous system**.

Somatic Nervous System

The somatic nervous system controls voluntary movements and transmits sensory information to the CNS. It consists of:

- **Sensory (afferent) nerves**, which carry information from sensory receptors (skin, muscles, joints) to the CNS.
- **Motor (efferent) nerves**, which deliver commands from the CNS to skeletal muscles.

For example, when you touch a hot surface, sensory nerves transmit the sensation to your brain, which processes the information and sends motor commands to withdraw your hand.

The somatic system also coordinates reflex arcs, which are automatic responses that bypass the brain. In a reflex arc, sensory neurons synapse directly with motor neurons in the spinal cord, enabling a rapid response.

Autonomic Nervous System (ANS)

The autonomic nervous system regulates involuntary functions, such as heartbeat, digestion, and respiratory rate. It operates below conscious control and is divided into three branches:

1. **Sympathetic Nervous System:** Known for the "fight or flight" response, it prepares the body for action during stress or danger. It increases heart rate, dilates airways, and redirects blood flow to muscles.
2. **Parasympathetic Nervous System:** Often called the "rest and digest" system, it conserves energy and promotes relaxation. It slows the heart rate, stimulates digestion, and facilitates nutrient absorption.
3. **Enteric Nervous System:** This subsystem governs the gastrointestinal tract. Sometimes called the "second brain," it controls digestion independently but communicates with the CNS through the sympathetic and parasympathetic systems.

Cranial and Spinal Nerves

The PNS includes 12 pairs of **cranial nerves** and 31 pairs of **spinal nerves**, each serving specific regions of the body.

1. **Cranial Nerves:** These nerves emerge directly from the brain and brainstem, innervating the head, neck, and some internal organs. Examples include:
    - **Olfactory nerve (I):** Responsible for the sense of smell.
    - **Optic nerve (II):** Transmits visual information from the eyes to the brain.
    - **Vagus nerve (X):** Regulates heart rate, digestion, and respiratory functions.

2. **Spinal Nerves:** Each spinal nerve emerges from the spinal cord through an intervertebral foramen and divides into dorsal (sensory) and ventral (motor) roots. Spinal nerves innervate specific regions, forming networks called **plexuses**. For example:
    - The **cervical plexus** supplies the neck and shoulders.
    - The **brachial plexus** controls the arms and hands.
    - The **lumbar and sacral plexuses** innervate the pelvis and legs.

Integration of CNS and PNS

The CNS and PNS work together seamlessly. The CNS processes incoming sensory information and formulates responses, while the PNS transmits signals to and from the CNS. For instance, when you see a ball coming toward you, sensory information from the optic nerve is processed in the CNS, and motor commands are sent via the PNS to move your arm and catch the ball.

The PNS also provides feedback to the CNS, allowing constant monitoring and adjustment. Sensory nerves report changes in the environment, and the CNS uses this information to adapt responses, maintaining balance, coordination, and homeostasis.

Neuroglia: Supporting Cells

While neurons are the primary functional units of the nervous system, **neuroglia** (glial cells) support and protect them. These cells differ in function between the CNS and PNS:

1. **CNS Glial Cells:**

    - **Astrocytes:** Maintain the blood-brain barrier, regulate nutrients, and repair damaged tissue.
    - **Oligodendrocytes:** Produce myelin, which insulates axons and speeds up signal transmission.
    - **Microglia:** Act as immune cells, removing debris and pathogens.
    - **Ependymal Cells:** Line ventricles and produce cerebrospinal fluid.

2. **PNS Glial Cells:**

    - **Schwann Cells:** Myelinate axons in the PNS, aiding in signal conduction and nerve regeneration.
    - **Satellite Cells:** Surround neuron cell bodies in ganglia, regulating their environment.

The interaction between neurons and glial cells ensures the nervous system functions efficiently, supporting both communication and repair.

How Neurons Communicate: Synapses and Neurotransmitters

Neurons communicate through specialized junctions called **synapses**, where signals are transmitted from one neuron to another or from a neuron to an effector cell, such as a muscle or gland. Communication occurs via electrical impulses and chemical messengers called **neurotransmitters**, which ensure the precise and rapid flow of information within the nervous system.

Structure of a Synapse

A synapse consists of three main components: the **presynaptic neuron**, the **synaptic cleft**, and the **postsynaptic cell**. The presynaptic neuron ends in a terminal button, containing synaptic vesicles filled with neurotransmitters. The synaptic cleft is the small gap between the two cells, and the postsynaptic cell contains receptors that bind to neurotransmitters, initiating a response.

Electrical to Chemical Transmission

When an action potential reaches the axon terminal of the presynaptic neuron, it triggers the opening of voltage-gated **calcium channels**. Calcium ions enter the terminal and stimulate the fusion of synaptic vesicles with the presynaptic membrane. This process, called **exocytosis**, releases neurotransmitters into the synaptic cleft.

The neurotransmitters diffuse across the cleft and bind to **specific receptors** on the postsynaptic membrane. The type of receptor and neurotransmitter determines the effect. For example, binding may open ion channels or activate second messenger systems, leading to either excitation or inhibition of the postsynaptic cell.

Excitatory and Inhibitory Signals

Neurotransmitters can have excitatory or inhibitory effects. **Excitatory neurotransmitters**, like glutamate, depolarize the postsynaptic membrane by opening sodium channels, bringing the membrane potential closer to the threshold for firing an action potential. In contrast, **inhibitory neurotransmitters**, like gamma-aminobutyric acid (GABA), hyperpolarize the membrane by opening potassium or chloride channels, making it harder for the cell to reach the threshold.

The balance between excitation and inhibition determines whether the postsynaptic cell generates an action potential. This balance is critical for proper nervous system function, as imbalances can lead to disorders like epilepsy or anxiety.

Types of Neurotransmitters

Neurotransmitters are diverse in structure and function. Some common types include:

- **Acetylcholine (ACh):** Found at neuromuscular junctions and in the autonomic nervous system, ACh is essential for muscle contraction and parasympathetic signaling.
- **Dopamine:** Involved in reward, motivation, and motor control, dopamine imbalances are linked to conditions like Parkinson's disease and schizophrenia.

- **Serotonin:** Regulates mood, appetite, and sleep. Low serotonin levels are associated with depression.
- **Norepinephrine:** A key player in the sympathetic nervous system, it enhances alertness and prepares the body for stress.
- **Glutamate:** The primary excitatory neurotransmitter in the CNS, crucial for learning and memory.
- **GABA:** The main inhibitory neurotransmitter in the brain, it reduces neural excitability and prevents overstimulation.

Synaptic Plasticity

Synapses are not static; they adapt based on activity. This **synaptic plasticity** underlies learning and memory. Strengthening a synapse, known as **long-term potentiation (LTP)**, occurs when repeated activity increases the postsynaptic cell's sensitivity to neurotransmitters. Conversely, **long-term depression (LTD)** reduces synaptic strength, refining neural circuits and eliminating unnecessary connections.

Neurotransmitter Clearance

After transmitting a signal, neurotransmitters must be cleared from the synaptic cleft to reset the system. This occurs through reuptake into the presynaptic neuron, enzymatic degradation, or diffusion away from the cleft. For example, acetylcholine is broken down by **acetylcholinesterase**, while serotonin and dopamine are reabsorbed by transport proteins.

Reflexes and the Autonomic Nervous System

Reflexes are rapid, automatic responses to stimuli that protect the body from harm and maintain homeostasis. The autonomic nervous system (ANS) operates involuntarily, regulating internal organs, blood vessels, and glands. Together, reflexes and the ANS ensure that the body responds quickly to internal and external changes.

Reflex Arcs

Reflexes are mediated by **reflex arcs**, neural pathways that bypass conscious brain activity to produce immediate responses. A reflex arc typically involves five components:

1. **Receptor:** Detects the stimulus, such as a pain receptor sensing a sharp object.
2. **Sensory Neuron:** Transmits the signal to the central nervous system.
3. **Integration Center:** Processes the information, often within the spinal cord.

4. **Motor Neuron:** Carries the response signal to the effector.
5. **Effector:** Executes the response, such as withdrawing a hand from a hot surface.

For example, in the **patellar reflex** (knee-jerk response), a tap on the patellar tendon stretches the quadriceps muscle. This activates stretch receptors, which send signals to the spinal cord. The integration center immediately sends a command via motor neurons to contract the quadriceps, causing the leg to kick.

Types of Reflexes

Reflexes can be classified as:

- **Somatic Reflexes:** Involve skeletal muscles and protect the body from injury. Examples include the withdrawal reflex and the corneal reflex (blinking in response to an object approaching the eye).
- **Autonomic Reflexes:** Regulate internal organ functions. For instance, the **baroreceptor reflex** adjusts blood pressure by altering heart rate and blood vessel diameter in response to changes in arterial pressure.

Overview of the Autonomic Nervous System

The ANS is divided into the **sympathetic**, **parasympathetic**, and **enteric** divisions, each with distinct roles in maintaining homeostasis.

1. **Sympathetic Division:** Prepares the body for emergencies through the "fight or flight" response. When activated, the sympathetic system increases heart rate, dilates airways, and redirects blood flow to muscles. This response is mediated by norepinephrine and epinephrine, which bind to adrenergic receptors in target tissues.

2. **Parasympathetic Division:** Promotes rest and recovery through the "rest and digest" response. It slows the heart rate, stimulates digestion, and conserves energy. Acetylcholine is the primary neurotransmitter in this division, acting on muscarinic receptors.

3. **Enteric Division:** Governs the gastrointestinal system independently but communicates with the CNS via the sympathetic and parasympathetic systems. It regulates processes like peristalsis, enzyme secretion, and nutrient absorption.

Control of the ANS

The ANS is regulated by the **hypothalamus**, which integrates sensory input and coordinates autonomic responses. For example, during dehydration, the hypothalamus stimulates the release of antidiuretic hormone (ADH) to conserve water and reduce urine output. Similarly, in response to stress, it activates the sympathetic division to prepare the body for action.

The brainstem also contributes to autonomic control. The **medulla oblongata** regulates heart rate, blood pressure, and respiratory rhythms, while the **pons** influences breathing patterns.

Reflexes in the ANS

Autonomic reflexes are critical for maintaining physiological stability. For instance:

- The **pupillary light reflex** controls the size of the pupil in response to light intensity. Bright light triggers the parasympathetic division to constrict the pupil, protecting the retina.
- The **defecation reflex** activates when the rectum is stretched, signaling the enteric nervous system to initiate bowel movements.
- The **vasovagal reflex** can cause fainting in response to stress or pain, as an overactive parasympathetic response slows the heart rate and dilates blood vessels.

Disorders of Reflexes and the ANS

Disruptions in reflex pathways or autonomic regulation can lead to various disorders. For example, damage to the spinal cord can impair reflexes and autonomic control below the injury level, causing issues like incontinence or blood pressure instability. Overactive reflexes in conditions like **spasticity** result in exaggerated muscle contractions, while diminished reflexes can indicate nerve damage or neurodegenerative diseases.

In the ANS, imbalances between sympathetic and parasympathetic activity contribute to disorders like hypertension, irritable bowel syndrome, and chronic stress. Treatments targeting specific receptors or neurotransmitters, such as beta-blockers or anticholinergics, can help restore balance and alleviate symptoms.

Neural Plasticity: How the Brain Adapts and Changes

Neural plasticity, also known as neuroplasticity, refers to the brain's ability to adapt and reorganize itself in response to changes in the environment, experiences, injury, or learning. This flexibility allows neurons to form new connections, strengthen existing ones, and even repurpose certain brain regions to compensate for damage or adapt to new challenges. Neural plasticity is fundamental to memory, learning, and recovery from brain injuries.

Types of Neural Plasticity

Neuroplasticity occurs in various forms, depending on the context and scale of changes:

1. **Structural Plasticity:** This involves physical changes in the brain, such as the growth of new dendrites, axon sprouting, or the formation of new synapses. For example, when learning a new skill like playing an instrument, the brain strengthens and reorganizes pathways involved in motor control and auditory processing.

2. **Functional Plasticity:** This type refers to the brain's ability to shift functions from damaged areas to healthy regions. For instance, after a stroke, undamaged areas of the brain can take over the lost functions through rewiring and adaptation.

3. **Synaptic Plasticity:** At the cellular level, synaptic plasticity involves the strengthening or weakening of synapses in response to activity. Synapses that are frequently used become stronger through **long-term potentiation (LTP)**, while those that are less active weaken, a process known as **long-term depression (LTD)**.

Mechanisms of Plasticity

Neuroplasticity depends on changes at both the cellular and molecular levels. These changes are driven by activity-dependent mechanisms, where neurons adjust their connections based on the frequency and intensity of their use.

- **Hebbian Plasticity:** Often summarized as "cells that fire together, wire together," Hebbian plasticity emphasizes that coordinated activity between two neurons strengthens their synaptic connection. This mechanism is critical for learning and memory formation.

- **Synaptogenesis and Pruning:** The brain is particularly plastic during early development, when synapses are overproduced and later pruned based on experience. Synaptogenesis creates new connections, while pruning removes unused ones, refining neural circuits.

- **Neurogenesis:** In certain regions, like the hippocampus, new neurons can be generated even in adulthood. This process supports memory and emotional regulation, although its extent and significance are still under study.

- **Chemical and Structural Changes:** Plasticity also involves alterations in neurotransmitter release, receptor density, and the cytoskeletal structure of neurons. For example, repeated activation of a synapse increases the number of AMPA receptors on the postsynaptic membrane, strengthening the connection.

Plasticity in Learning and Memory

Learning and memory rely heavily on neural plasticity. When a person learns a new skill or piece of information, neural circuits involved in processing and storing that information are reshaped. **Explicit memory**, which involves facts and events, primarily engages the hippocampus and medial temporal lobes. During learning, LTP strengthens the synaptic connections in these regions, ensuring that the memory is consolidated.

Motor learning, such as acquiring a new physical skill, relies on changes in the motor cortex and cerebellum. Repeated practice refines these neural pathways, increasing their efficiency. For instance, pianists have highly developed motor areas corresponding to finger movements due to extensive practice.

Plasticity After Injury

Neuroplasticity is a critical factor in recovery from brain injuries such as strokes, traumatic brain injuries, or neurodegenerative diseases. After an injury, the brain attempts to reorganize itself by rerouting functions to undamaged areas or recruiting alternate pathways. For example, in people who lose the ability to speak due to damage in **Broca's area**, nearby regions may take on some language processing tasks with therapy and practice.

However, recovery depends on the extent of the injury, age, and rehabilitation efforts. Young brains exhibit greater plasticity, allowing children to recover from injuries more effectively than adults. Rehabilitation strategies, such as physical therapy and cognitive exercises, leverage neuroplasticity to enhance recovery.

Experience-Dependent Plasticity

The brain is shaped by experiences throughout life. Enriched environments, characterized by sensory stimulation, physical activity, and social interaction, promote the growth of dendrites and synapses. For example, children exposed to diverse stimuli during critical developmental periods develop stronger and more complex neural networks.

Conversely, deprivation or chronic stress can impair plasticity. Prolonged stress elevates cortisol levels, which can damage the hippocampus, reducing its capacity for memory and learning. Similarly, neglect during early development can limit synaptic growth, resulting in long-term cognitive deficits.

Technological and Therapeutic Applications

Neuroplasticity has inspired numerous therapeutic approaches and technologies aimed at enhancing brain function or recovery. For instance:

- **Cognitive Behavioral Therapy (CBT):** CBT relies on the brain's plasticity to change thought patterns and behaviors. By challenging negative beliefs, the therapy helps rewire neural circuits associated with emotions and decision-making.

- **Brain-Computer Interfaces (BCIs):** BCIs leverage neuroplasticity to help individuals with paralysis control devices through neural signals. Training with BCIs strengthens the relevant neural pathways, improving performance over time.

- **Transcranial Magnetic Stimulation (TMS):** TMS uses magnetic fields to stimulate specific brain regions, promoting plasticity and aiding in the treatment of depression and other disorders.

- **Rehabilitation Robotics:** Robots designed for physical therapy help patients relearn movements, enhancing neural plasticity through repetitive and guided practice.

Limits and Challenges

While neuroplasticity offers remarkable adaptability, it has limitations. In some cases, plasticity can lead to maladaptive changes. For example, **chronic pain** involves plasticity in pain-processing circuits, amplifying the perception of pain even after an injury has healed. Similarly, **phantom limb syndrome** occurs when the brain reorganizes itself after limb loss, leading to the perception of sensations in the missing limb.

Plasticity also declines with age, reducing the brain's ability to adapt. However, lifelong learning and physical activity can mitigate these effects, preserving cognitive function and encouraging continued plasticity.

Neuroplasticity demonstrates the nervous system's ability to adapt and thrive in a changing environment. Whether in response to learning, injury, or experience, the brain's capacity for change underscores its remarkable resilience and complexity.

CHAPTER 9: THE ENDOCRINE SYSTEM

Glands and Hormones: Key Players in Regulation

The endocrine system regulates the body's internal environment by releasing hormones, chemical messengers that travel through the bloodstream to target organs. These hormones control growth, metabolism, reproduction, and many other functions. The glands of the endocrine system secrete hormones in precise amounts, ensuring that physiological processes remain balanced.

Hypothalamus: The Master Controller

The **hypothalamus** connects the nervous system to the endocrine system. Located in the brain, it regulates hormone secretion by controlling the activity of the **pituitary gland**. The hypothalamus produces **releasing hormones** and **inhibiting hormones**, which stimulate or suppress the release of hormones from the anterior pituitary.

In addition to regulating the pituitary gland, the hypothalamus produces two hormones stored and released by the posterior pituitary: **oxytocin**, which facilitates childbirth and lactation, and **antidiuretic hormone (ADH)**, which regulates water balance by influencing kidney function.

Pituitary Gland: The Central Regulator

The **pituitary gland**, often called the "master gland," secretes hormones that influence many other endocrine glands. It is divided into two parts: the anterior pituitary and the posterior pituitary.

The **anterior pituitary** produces hormones that regulate growth, reproduction, and metabolism:

- **Growth hormone (GH):** Stimulates growth of bones, muscles, and tissues. GH also regulates metabolism by promoting fat breakdown and glucose production.
- **Adrenocorticotropic hormone (ACTH):** Stimulates the adrenal cortex to release cortisol, which helps the body respond to stress and regulate metabolism.
- **Thyroid-stimulating hormone (TSH):** Signals the thyroid gland to produce thyroid hormones, which control metabolism.
- **Prolactin (PRL):** Promotes milk production in lactating women.

- **Follicle-stimulating hormone (FSH) and luteinizing hormone (LH):** Regulate the reproductive system by controlling the function of ovaries in females and testes in males.

The **posterior pituitary** does not produce hormones but stores and releases oxytocin and ADH produced by the hypothalamus.

Thyroid Gland: Regulating Metabolism

The **thyroid gland**, located in the neck, produces hormones that regulate metabolism, growth, and development. The primary hormones secreted by the thyroid are **thyroxine (T4)** and **triiodothyronine (T3)**. These hormones increase the rate at which cells convert nutrients into energy, influencing metabolic rate, heart rate, and body temperature.

Thyroid function is tightly controlled by TSH from the pituitary gland. When T3 and T4 levels drop, the pituitary releases more TSH to stimulate the thyroid. A deficiency in thyroid hormones can lead to **hypothyroidism**, characterized by fatigue, weight gain, and cold intolerance. Excess thyroid hormone causes **hyperthyroidism**, leading to weight loss, increased heart rate, and heat sensitivity.

The thyroid also produces **calcitonin**, a hormone that lowers blood calcium levels by inhibiting bone resorption and promoting calcium storage in bones.

Parathyroid Glands: Calcium Balance

The **parathyroid glands**, located on the posterior surface of the thyroid, produce **parathyroid hormone (PTH)**. PTH raises blood calcium levels by stimulating bone resorption, increasing calcium absorption in the intestines, and reducing calcium excretion by the kidneys. PTH works in opposition to calcitonin to maintain calcium homeostasis, which is vital for muscle contraction, nerve function, and blood clotting.

Adrenal Glands: Responding to Stress

The **adrenal glands**, located on top of the kidneys, consist of two distinct parts: the adrenal cortex and the adrenal medulla. Each part produces hormones with different functions.

The **adrenal cortex** produces:

- **Cortisol:** Helps the body respond to stress by increasing glucose availability, reducing inflammation, and influencing metabolism. Cortisol secretion is regulated by ACTH from the pituitary.
- **Aldosterone:** Maintains blood pressure and electrolyte balance by promoting sodium retention and potassium excretion in the kidneys.

- **Androgens:** Weak male sex hormones that are converted into more potent forms in other tissues.

The **adrenal medulla**, the inner portion of the gland, produces **epinephrine (adrenaline)** and **norepinephrine (noradrenaline)**. These hormones are released during the "fight or flight" response, increasing heart rate, blood pressure, and blood flow to muscles while reducing digestion and other non-essential functions.

Pancreas: Glucose Regulation

The **pancreas** serves both endocrine and exocrine functions. Its endocrine role involves regulating blood glucose levels through the secretion of two key hormones:

- **Insulin:** Lowers blood glucose by promoting the uptake of glucose into cells and stimulating glycogen storage in the liver. Insulin is released by beta cells in the pancreas when blood sugar levels are high.
- **Glucagon:** Raises blood glucose by stimulating glycogen breakdown in the liver and promoting glucose release into the bloodstream. Glucagon is secreted by alpha cells when blood sugar levels are low.

Disruption in insulin production or function leads to **diabetes mellitus**, a condition characterized by high blood sugar levels. Type 1 diabetes occurs when the pancreas produces little or no insulin, while Type 2 diabetes involves insulin resistance.

Pineal Gland: Regulating Sleep

The **pineal gland**, a small structure in the brain, secretes **melatonin**, which regulates sleep-wake cycles. Melatonin production increases in response to darkness and decreases in light, aligning the body's internal clock with external cues. This hormone influences circadian rhythms and helps the body prepare for rest.

Reproductive Glands: Hormones and Fertility

The **ovaries** in females and the **testes** in males produce hormones essential for reproduction and secondary sexual characteristics.

The **ovaries** secrete:

- **Estrogen:** Promotes the development of female secondary sexual characteristics, regulates the menstrual cycle, and supports pregnancy.
- **Progesterone:** Prepares the uterus for implantation of a fertilized egg and supports pregnancy.

The **testes** produce:

- **Testosterone:** Stimulates the development of male secondary sexual characteristics, supports sperm production, and promotes muscle and bone growth.

FSH and LH from the pituitary regulate the function of these glands, ensuring proper hormonal balance and reproductive function.

Hormonal Regulation and Feedback Loops

The endocrine system operates through feedback loops to maintain balance. Most hormones are regulated by **negative feedback**, where rising levels of a hormone suppress further secretion. For example, high levels of thyroid hormones inhibit TSH release, preventing overproduction.

In some cases, **positive feedback** amplifies a process. For instance, during childbirth, oxytocin release causes uterine contractions, which stimulate further oxytocin release until delivery is complete.

Disorders of the Endocrine System

Imbalances in hormones can lead to endocrine disorders. **Hypersecretion** of hormones, such as cortisol in **Cushing's syndrome**, causes symptoms like weight gain and high blood pressure. **Hyposecretion**, as seen in **Addison's disease**, results in fatigue, low blood pressure, and electrolyte imbalances. Early detection and treatment, often involving hormone replacement or medication, are critical for managing these conditions.

The Role of the Endocrine System in Growth, Metabolism, and Reproduction

The endocrine system orchestrates complex processes like growth, metabolism, and reproduction by releasing hormones into the bloodstream. These hormones interact with specific receptors in target tissues, triggering physiological changes essential for development, energy balance, and fertility.

Growth and Development

Growth is primarily regulated by **growth hormone (GH)**, which is secreted by the anterior pituitary gland. GH promotes the growth of bones, muscles, and other tissues by stimulating protein synthesis and cell division. It acts on the liver to produce **insulin-like growth factor-1 (IGF-1)**, which further stimulates the growth of cartilage in the epiphyseal plates of long bones during childhood and adolescence.

Thyroid hormones, specifically **thyroxine (T4)** and **triiodothyronine (T3)**, are also crucial for normal growth and development. They regulate the growth of the nervous system in the fetal and early postnatal periods. Deficiencies in thyroid hormones during these stages can lead to developmental delays or conditions like cretinism.

Sex hormones, including **testosterone** and **estrogen**, contribute to the growth spurt and skeletal maturation during puberty. Testosterone stimulates muscle growth and bone density, while estrogen promotes the closure of epiphyseal plates, marking the end of height growth.

Parathyroid hormone (PTH) and **calcitriol** (active vitamin D) regulate calcium levels, supporting bone mineralization and strength. Without proper calcium balance, growth and skeletal health are compromised, increasing the risk of conditions like rickets in children or osteoporosis in adults.

Metabolism and Energy Regulation

Metabolism, the process by which the body converts food into energy, is tightly controlled by hormones from the thyroid, pancreas, and adrenal glands. Thyroid hormones are central to regulating basal metabolic rate (BMR). They increase oxygen consumption and heat production by stimulating mitochondrial activity in cells. An overactive thyroid (hyperthyroidism) accelerates metabolism, causing weight loss and increased energy expenditure, while an underactive thyroid (hypothyroidism) slows metabolism, leading to weight gain and fatigue.

The pancreas regulates blood glucose levels through **insulin** and **glucagon**. Insulin, secreted by beta cells, promotes glucose uptake by cells and its storage as glycogen in the liver. Glucagon, released by alpha cells, stimulates glycogen breakdown and glucose release when blood sugar levels are low. Together, these hormones maintain energy homeostasis.

The adrenal glands also influence metabolism through cortisol, which modulates glucose production, fat metabolism, and protein breakdown during stress. Cortisol ensures that energy is available to tissues during prolonged physical or psychological challenges.

Reproduction and Fertility

Reproductive functions depend on the precise regulation of hormones from the hypothalamus, pituitary gland, and gonads. The hypothalamus releases **gonadotropin-releasing hormone (GnRH)**, which stimulates the pituitary to secrete **follicle-stimulating hormone (FSH)** and **luteinizing hormone (LH)**.

In females, FSH promotes the maturation of ovarian follicles, while LH triggers ovulation and the formation of the **corpus luteum**, which secretes progesterone to prepare the uterus for implantation. Estrogen and progesterone regulate the menstrual cycle and maintain pregnancy by promoting uterine growth and preventing contractions.

In males, FSH and LH stimulate the testes to produce sperm and testosterone. Testosterone influences sperm maturation, libido, and the development of secondary sexual characteristics like deeper voice and facial hair.

Prolactin, another hormone from the anterior pituitary, stimulates milk production in lactating women. Oxytocin, released by the posterior pituitary, facilitates milk ejection and strengthens uterine contractions during labor.

Hormonal imbalances can disrupt reproductive functions, leading to infertility or disorders like polycystic ovary syndrome (PCOS) in women and low testosterone or sperm count in men.

Feedback Mechanisms and Hormonal Balance

The endocrine system relies on feedback mechanisms to maintain hormonal balance and ensure physiological processes remain within normal ranges. These mechanisms, primarily **negative feedback loops**, regulate hormone secretion by modulating the activity of glands in response to changes in the internal environment.

Negative Feedback Loops

Negative feedback prevents overproduction or underproduction of hormones. For instance, the hypothalamic-pituitary-thyroid (HPT) axis controls thyroid hormone levels. When blood levels of thyroxine (T4) and triiodothyronine (T3) rise, they inhibit the release of **thyroid-stimulating hormone (TSH)** from the pituitary and **thyrotropin-releasing hormone (TRH)** from the hypothalamus. This reduces thyroid hormone production, maintaining a stable concentration.

A similar loop operates in the hypothalamic-pituitary-adrenal (HPA) axis. Cortisol, secreted by the adrenal cortex, suppresses the release of **adrenocorticotropic hormone (ACTH)** and **corticotropin-releasing hormone (CRH)** when its levels are high, preventing excessive stress responses and metabolic disruption.

Insulin and glucagon also function via negative feedback to stabilize blood glucose levels. High blood sugar triggers insulin secretion, lowering glucose by promoting cellular uptake and storage. When blood sugar drops, glucagon is released to raise glucose levels through glycogen breakdown and gluconeogenesis.

Positive Feedback Loops

While rare, positive feedback amplifies a process until a specific outcome is achieved. A well-known example is the release of oxytocin during childbirth. Uterine contractions stimulate oxytocin secretion, which intensifies the contractions, further increasing oxytocin release. This cycle continues until delivery, after which the feedback loop stops.

Another example occurs during the menstrual cycle's follicular phase. Rising estrogen levels stimulate the release of LH in a positive feedback loop, leading to the LH surge that triggers ovulation. Once ovulation occurs, the feedback shifts to negative to regulate hormone levels.

Hormonal Rhythms and Regulation

Many hormones follow daily or seasonal rhythms, ensuring their effects align with the body's needs. **Cortisol**, for example, peaks in the morning to prepare the body for activity and gradually declines throughout the day. Disruptions in these rhythms, such as from shift work or chronic stress, can lead to hormonal imbalances and health issues like insomnia or metabolic disorders.

The **pineal gland** regulates melatonin secretion based on light exposure, synchronizing sleep-wake cycles with environmental cues. Hormonal rhythms are essential for maintaining energy balance, reproductive cycles, and overall homeostasis.

Impact of Dysregulation

Hormonal imbalances can disrupt feedback mechanisms, leading to endocrine disorders. For example, in **Graves' disease**, an autoimmune condition, antibodies stimulate the thyroid gland independently of TSH, causing excessive thyroid hormone production and hyperthyroidism. Conversely, in **Hashimoto's thyroiditis**, the immune system damages the thyroid, leading to reduced hormone output and hypothyroidism.

Diabetes mellitus exemplifies the disruption of feedback in glucose regulation. In Type 1 diabetes, insulin production is impaired, while in Type 2 diabetes, insulin resistance prevents effective glucose uptake. Both conditions result in chronically high blood sugar levels, requiring medical intervention to restore balance.

Therapeutic Applications

Understanding feedback mechanisms allows for targeted treatments. Synthetic hormones, such as levothyroxine for hypothyroidism or insulin for diabetes, supplement deficient hormones, restoring normal function. Medications like beta-blockers or aldosterone antagonists modulate excessive hormonal effects, improving outcomes in conditions like hyperthyroidism or hypertension.

Modern technologies, such as continuous glucose monitors and automated insulin pumps, leverage feedback principles to maintain glucose control in diabetic patients. Similarly, hormone replacement therapies help manage imbalances during menopause or adrenal insufficiency.

Endocrine Disruptors and Their Impact on Health

Endocrine disruptors are chemicals that interfere with the normal functioning of the endocrine system. These substances can mimic, block, or alter the production, release, transport, or action of hormones, leading to imbalances that affect growth, reproduction, metabolism, and other critical processes. Found in everyday products and the environment, endocrine disruptors have become a significant concern for human and environmental health.

Sources and Types of Endocrine Disruptors

Endocrine disruptors are present in a wide range of materials, including plastics, pesticides, industrial chemicals, and personal care products. Some common examples include:

- **Bisphenol A (BPA):** Found in polycarbonate plastics and epoxy resins used in food and beverage containers, BPA mimics estrogen and can bind to estrogen receptors, disrupting normal hormonal activity.
- **Phthalates:** Used as plasticizers in vinyl products and as solvents in cosmetics, phthalates interfere with testosterone production and are associated with reproductive issues.
- **Polychlorinated biphenyls (PCBs):** Industrial chemicals banned in many countries but still persistent in the environment, PCBs disrupt thyroid hormone function and are linked to neurodevelopmental disorders.
- **Dioxins:** Byproducts of industrial processes like waste incineration, dioxins affect reproductive and immune systems by altering estrogen and androgen signaling.
- **Pesticides:** Chemicals like DDT and atrazine disrupt hormone pathways by mimicking or blocking estrogen and other hormones, affecting fertility and development.

Natural substances, such as **phytoestrogens** in soy, can also act as endocrine disruptors, but their effects are often less pronounced and context-dependent.

Mechanisms of Disruption

Endocrine disruptors can interfere with hormonal pathways through several mechanisms:

1. **Mimicking Hormones:** Some disruptors, such as BPA, act as hormone mimics, binding to receptors intended for natural hormones like estrogen or androgen. This inappropriate activation can overstimulate cellular responses or disrupt normal feedback loops.

2. **Blocking Hormone Receptors:** Disruptors like certain pesticides prevent natural hormones from binding to their receptors. For example, anti-androgenic chemicals block testosterone action, impairing male reproductive development.

3. **Altering Hormone Synthesis and Metabolism:** Endocrine disruptors can affect enzymes involved in hormone production or breakdown. For instance, some disruptors inhibit the synthesis of thyroid hormones, leading to imbalances that affect metabolism and brain development.

4. **Interfering with Hormone Transport:** Hormones rely on carrier proteins to travel through the bloodstream. Disruptors can bind to these proteins, displacing natural hormones and reducing their availability.

Health Impacts of Endocrine Disruptors

The health effects of endocrine disruptors depend on the timing, dose, and duration of exposure. Developing fetuses, infants, and children are particularly vulnerable due to the rapid growth and differentiation of their organs and tissues.

- **Reproductive Disorders:** Exposure to endocrine disruptors during critical developmental windows can lead to infertility, altered sexual development, and other reproductive problems. In males, chemicals like phthalates are linked to reduced sperm quality, testicular dysgenesis, and undescended testes. In females, disruptors such as BPA may contribute to polycystic ovary syndrome (PCOS), early puberty, and impaired fertility.

- **Developmental and Neurobehavioral Effects:** Endocrine disruptors like PCBs and dioxins impact brain development, leading to learning disabilities, attention deficits, and reduced IQ. These effects often result from disruptions in thyroid hormone signaling, which is critical for neurodevelopment.

- **Metabolic Disorders:** Some disruptors, termed **obesogens**, contribute to weight gain and metabolic dysfunction by altering fat storage, appetite regulation, and insulin sensitivity. For example, BPA exposure is associated with increased risk of obesity, type 2 diabetes, and metabolic syndrome.

- **Cancer Risk:** Endocrine disruptors that mimic estrogen, such as BPA and DDT, are linked to hormone-dependent cancers, including breast, ovarian, and prostate cancers. These chemicals promote cell proliferation and disrupt the natural regulation of cell growth.

- **Immune System Dysfunction:** Chemicals like dioxins suppress immune responses, increasing susceptibility to infections and reducing the effectiveness of vaccines.

Environmental and Generational Effects

The impact of endocrine disruptors extends beyond individual health, affecting ecosystems and future generations. Wildlife studies have shown that exposure to chemicals like DDT and PCBs causes reproductive abnormalities, such as feminization of male fish and birds, and reduced fertility in amphibians.

In humans, epigenetic changes induced by endocrine disruptors can be passed to offspring, amplifying health effects across generations. For instance, prenatal exposure to disruptors has been linked to altered gene expression in sperm and eggs, increasing the risk of metabolic, reproductive, and neurological disorders in subsequent generations.

Regulation and Mitigation

Efforts to minimize the impact of endocrine disruptors include stricter regulations and public awareness campaigns. Many countries have banned or restricted certain chemicals, such as PCBs and DDT. However, newer chemicals with similar effects continue to emerge, posing challenges for regulators.

Individuals can reduce exposure by avoiding products containing BPA or phthalates, choosing fresh or frozen foods instead of canned goods, and using natural cleaning and personal care products. Proper disposal of industrial chemicals and reduced reliance on pesticides are also critical for protecting the environment.

Advances in research, such as high-throughput screening and computational modeling, are helping identify endocrine-disrupting properties in chemicals more quickly. These tools aid in developing safer alternatives and improving regulatory frameworks.

Endocrine disruptors challenge the delicate balance maintained by the endocrine system, underscoring the need for vigilance in reducing their presence and mitigating their effects on health.

CHAPTER 10: THE CARDIOVASCULAR SYSTEM

Anatomy of the Heart and Blood Vessels

The heart and blood vessels form the cardiovascular system, responsible for circulating blood throughout the body. This system delivers oxygen and nutrients to tissues, removes waste products, and maintains homeostasis. Understanding the anatomy of the heart and blood vessels reveals how this complex network sustains life through its constant activity.

The Heart: Structure and Function

The **heart** is a muscular organ located in the thoracic cavity between the lungs, slightly tilted to the left. About the size of a fist, it weighs roughly 300 grams in adults. The heart is divided into four chambers: two atria (upper chambers) and two ventricles (lower chambers). These chambers are separated by the **septum**, a wall of tissue that prevents mixing of oxygen-rich and oxygen-poor blood.

The **right atrium** receives deoxygenated blood from the body through the **superior vena cava** and **inferior vena cava**. This blood flows into the **right ventricle**, which pumps it to the lungs via the **pulmonary artery** for oxygenation. The **left atrium** receives oxygenated blood from the lungs through the **pulmonary veins**. This blood is then delivered to the **left ventricle**, which pumps it to the rest of the body through the **aorta**, the largest artery.

The heart is enclosed by the **pericardium**, a double-layered sac that protects and anchors it. Between these layers is pericardial fluid, which reduces friction during heartbeats. The heart wall itself has three layers:

1. **Epicardium:** The outermost layer, which also forms part of the pericardium.
2. **Myocardium:** The thick middle layer composed of cardiac muscle, responsible for contraction and pumping action.
3. **Endocardium:** The smooth inner lining of the heart chambers and valves, which reduces resistance to blood flow.

Valves of the Heart

The heart has four valves that regulate blood flow, ensuring it moves in one direction. These valves open and close in response to pressure changes during the cardiac cycle:

- **Tricuspid Valve:** Located between the right atrium and right ventricle, it prevents backflow of blood into the atrium.
- **Pulmonary Valve:** Positioned at the exit of the right ventricle, it prevents blood from returning after it is pumped into the pulmonary artery.
- **Mitral Valve:** Found between the left atrium and left ventricle, it prevents backflow into the atrium.
- **Aortic Valve:** Located at the exit of the left ventricle, it stops blood from flowing back after it enters the aorta.

The rhythmic opening and closing of these valves produce the **"lub-dub"** sounds heard during a heartbeat.

The Cardiac Cycle

The heart functions through a repeating cycle of contraction and relaxation, known as the **cardiac cycle**:

- **Systole:** The ventricles contract, pumping blood into the pulmonary artery and aorta.
- **Diastole:** The ventricles relax, allowing them to fill with blood from the atria.

This cycle is driven by the heart's **electrical conduction system**, which ensures coordinated contractions. The **sinoatrial (SA) node**, located in the right atrium, acts as the natural pacemaker, generating electrical impulses that spread through the atria. These impulses reach the **atrioventricular (AV) node**, which delays the signal before passing it to the ventricles via the **bundle of His** and **Purkinje fibers**, ensuring efficient pumping.

Blood Vessels: Pathways for Circulation

Blood vessels form a vast network that transports blood throughout the body. They are categorized into three main types: **arteries**, **veins**, and **capillaries**. Each type is specialized for its role in circulation.

Arteries

Arteries carry oxygenated blood away from the heart to the body, with the exception of the pulmonary arteries, which transport deoxygenated blood to the lungs. Arteries have thick, elastic walls composed of three layers:

1. **Tunica Intima:** The innermost layer, lined with smooth endothelial cells to reduce friction.

2. **Tunica Media:** The middle layer of smooth muscle and elastic fibers, allowing arteries to withstand high pressure and regulate blood flow by constricting or dilating.
3. **Tunica Externa:** The outer layer of connective tissue that provides structural support.

The **aorta** branches into smaller arteries, which further divide into **arterioles**. Arterioles regulate blood flow to capillaries through vasodilation and vasoconstriction, controlled by the autonomic nervous system and local chemical signals.

Capillaries

Capillaries are the smallest blood vessels, forming extensive networks in tissues. Their thin walls, composed of a single layer of endothelial cells, allow efficient exchange of gases, nutrients, and waste products between blood and cells. Oxygen and nutrients diffuse from the blood into surrounding tissues, while carbon dioxide and metabolic waste move into the blood for removal.

Capillaries connect arterioles to **venules**, which transport deoxygenated blood back to the veins. The density of capillaries in a tissue depends on its metabolic activity; for example, muscles and organs like the liver have extensive capillary networks.

Veins

Veins return deoxygenated blood to the heart, except for the pulmonary veins, which carry oxygenated blood from the lungs. Veins have thinner walls than arteries but a larger diameter, allowing them to hold more blood. Like arteries, veins have three layers, but the tunica media is less muscular and elastic.

To prevent backflow, especially in the lower body, veins have **valves** that ensure blood flows toward the heart. Skeletal muscle contractions and pressure changes during breathing assist in venous return, a mechanism known as the **muscle pump**.

The largest veins in the body, the **superior vena cava** and **inferior vena cava**, drain blood from the upper and lower parts of the body into the right atrium.

Coronary Circulation

The heart has its own blood supply, known as **coronary circulation**, to meet its high metabolic demands. The **coronary arteries** branch off the aorta, delivering oxygenated blood to the myocardium. The **left coronary artery** supplies the left side of the heart, while the **right coronary artery** serves the right side. Deoxygenated blood is drained by the **coronary veins** into the **coronary sinus**, which empties into the right atrium.

Blockages in coronary arteries, often caused by atherosclerosis, can lead to **myocardial infarction (heart attack)** by depriving heart tissue of oxygen.

The Circulatory Pathway: Systemic and Pulmonary Circulation

The circulatory system operates through two interconnected pathways: **systemic circulation** and **pulmonary circulation**. These circuits ensure the continuous flow of blood, delivering oxygen and nutrients to tissues while removing carbon dioxide and waste. Together, they maintain the body's homeostasis by efficiently linking the heart, lungs, and body tissues.

Pulmonary Circulation

Pulmonary circulation is responsible for oxygenating blood. It begins in the **right ventricle**, where deoxygenated blood is pumped into the **pulmonary trunk** and subsequently into the **right and left pulmonary arteries**. These arteries transport the blood to the lungs, the only arteries in the body that carry deoxygenated blood.

In the lungs, blood travels through an extensive network of **pulmonary capillaries** surrounding the alveoli. Here, gas exchange occurs. Carbon dioxide diffuses from the blood into the alveoli to be exhaled, while oxygen enters the blood. The oxygenated blood then flows into the **pulmonary veins**, which return it to the **left atrium** of the heart. Pulmonary veins are unique because they are the only veins in the body that carry oxygen-rich blood.

Pulmonary circulation is relatively short, ensuring that blood is oxygenated quickly. This oxygen-rich blood then enters the systemic circuit for distribution throughout the body.

Systemic Circulation

Systemic circulation begins in the **left ventricle**, where oxygenated blood is pumped into the **aorta**, the body's largest artery. From the aorta, blood flows into major arteries that branch to supply specific regions. For example:

- The **carotid arteries** supply the brain and head.
- The **subclavian arteries** provide blood to the arms.
- The **thoracic and abdominal aorta** branch into arteries that serve the chest, abdomen, and lower limbs.

As arteries branch into smaller **arterioles** and eventually **capillaries**, the oxygen and nutrients carried in the blood are delivered to tissues. Simultaneously, carbon dioxide and waste products enter the blood, preparing it for removal.

Deoxygenated blood then flows from capillaries into **venules**, which merge into larger **veins**. The **superior vena cava** collects blood from the head and upper body, while the **inferior vena cava** gathers blood from the lower body. Both venae cavae empty into the **right atrium**, completing the systemic circuit. From there, the blood is sent back to the lungs via the pulmonary circuit to be reoxygenated.

Systemic circulation is the longer and higher-pressure circuit, as it must deliver blood to the entire body. The heart's left ventricle, with its thicker muscular wall, generates the force necessary for this extensive distribution.

Specialized Circulatory Routes

Systemic circulation includes several specialized pathways to address the needs of specific organs:

- **Hepatic Portal Circulation:** Blood from the digestive organs passes through the liver via the **hepatic portal vein** before returning to the heart. This allows the liver to metabolize nutrients, detoxify substances, and store glucose as glycogen.
- **Renal Circulation:** Blood flows through the kidneys to filter waste and regulate electrolyte balance. The kidneys receive a significant portion of cardiac output due to their vital role in homeostasis.
- **Coronary Circulation:** The heart itself is supplied with oxygen-rich blood through the coronary arteries. This circulation ensures the myocardium remains functional and capable of pumping blood.

The coordination of systemic and pulmonary circulation ensures that oxygen delivery and waste removal occur efficiently, meeting the demands of all body tissues.

Blood Composition and Its Functions

Blood is a specialized connective tissue that circulates through the cardiovascular system, transporting essential substances, defending against disease, and maintaining homeostasis. It is composed of plasma and cellular elements, each contributing to its diverse functions.

Plasma: The Liquid Component

Plasma makes up about 55% of blood volume and serves as the medium for transporting cells and dissolved substances. It is approximately 90% water, which helps maintain blood's fluidity and regulates body temperature. The remaining 10% consists of:

- **Proteins:** These include **albumin**, which maintains osmotic pressure and prevents fluid loss from blood vessels; **globulins**, which include antibodies for immune defense; and **fibrinogen**, essential for blood clotting.
- **Electrolytes:** Sodium, potassium, calcium, and chloride ions help regulate nerve function, muscle contractions, and pH balance.
- **Nutrients and Waste Products:** Plasma carries glucose, amino acids, lipids, and vitamins to tissues, while transporting urea, creatinine, and other waste to the kidneys for excretion.
- **Hormones and Gases:** Plasma distributes hormones from endocrine glands and transports oxygen, carbon dioxide, and nitrogen.

Plasma acts as the circulatory system's carrier, ensuring that cells and tissues receive the substances they need to function while removing byproducts for elimination.

Cellular Components of Blood

The cellular portion of blood accounts for about 45% of its volume and includes red blood cells, white blood cells, and platelets.

1. **Red Blood Cells (Erythrocytes):** These biconcave cells are responsible for oxygen transport. Each erythrocyte contains millions of molecules of **hemoglobin**, a protein that binds oxygen in the lungs and releases it in tissues. Hemoglobin also carries carbon dioxide back to the lungs for exhalation. The shape of red blood cells increases their surface area, enhancing oxygen exchange.

Red blood cells are produced in the **bone marrow** and have a lifespan of about 120 days. They are broken down in the spleen and liver, where their components are recycled.

2. **White Blood Cells (Leukocytes):** White blood cells defend the body against infections, foreign invaders, and abnormal cells. There are several types of leukocytes, each with a specific function:
    - **Neutrophils**: Engulf and destroy bacteria.
    - **Lymphocytes**: Include B cells that produce antibodies and T cells that destroy infected or cancerous cells.
    - **Monocytes**: Develop into macrophages, which remove debris and pathogens.
    - **Eosinophils and Basophils**: Play roles in allergic reactions and parasitic infections.

Leukocytes are fewer in number than red blood cells but are critical for immune function.

3. **Platelets (Thrombocytes):** Platelets are small cell fragments that aid in blood clotting. When a blood vessel is damaged, platelets adhere to the site,

forming a plug. They also release chemicals that activate the coagulation cascade, leading to the formation of a **fibrin clot** that seals the wound.

Functions of Blood

Blood performs several vital functions:

- **Transport:** Blood carries oxygen from the lungs to tissues, carbon dioxide from tissues to the lungs, nutrients from the digestive tract to cells, and waste products to the kidneys and liver for excretion.
- **Regulation:** Blood helps maintain pH balance through buffers, regulates body temperature by distributing heat, and controls fluid volume through osmotic pressure.
- **Protection:** White blood cells and antibodies in blood defend against pathogens, while platelets and clotting factors prevent excessive blood loss after injury.

Blood's composition and functions are intricately linked to the body's overall health. Its ability to adapt and respond to changing conditions ensures that all organs receive the support they need to function properly.

The Lymphatic System: Its Role in Circulation and Immunity

The **lymphatic system** works alongside the cardiovascular system to maintain fluid balance, transport dietary fats, and support immune defense. While blood vessels circulate oxygen and nutrients, the lymphatic system removes excess fluid, proteins, and waste products from tissues, returning them to the bloodstream. At the same time, it monitors the body for pathogens and facilitates immune responses, serving as a crucial link between circulation and immunity.

Structure of the Lymphatic System

The lymphatic system consists of a network of **lymphatic vessels**, **lymph nodes**, and associated organs, including the spleen, thymus, and tonsils. Lymph, a clear fluid derived from interstitial fluid, flows through this system, carrying waste, immune cells, and other materials.

1. **Lymphatic Vessels:** These vessels begin as blind-ended **lymphatic capillaries** in tissues. Their thin walls and overlapping endothelial cells allow them to absorb excess interstitial fluid, large proteins, and even pathogens. Unlike blood vessels, lymphatic capillaries do not form a closed loop. Lymphatic capillaries merge into larger **lymphatic vessels**, which are similar to veins but have thinner walls and more valves to prevent backflow.

These vessels transport lymph toward the thoracic region, where it is returned to the bloodstream through two major ducts:

- The **thoracic duct**, which drains lymph from most of the body into the **left subclavian vein**.
- The **right lymphatic duct**, which drains lymph from the upper right quadrant of the body into the **right subclavian vein**.

2. **Lymph Nodes:** Along the lymphatic vessels are **lymph nodes**, small, bean-shaped structures that filter lymph. Each node contains immune cells, such as lymphocytes and macrophages, that trap and destroy pathogens, cancer cells, and debris. Lymph nodes are concentrated in areas like the neck, armpits, and groin, where they act as checkpoints for monitoring lymph before it reenters the bloodstream.

3. **Lymphoid Organs:**
   - The **spleen** filters blood, removing old or damaged red blood cells and storing immune cells. It also acts as a reservoir for platelets and monocytes.
   - The **thymus** is critical for the maturation of T lymphocytes (T cells), which are essential for adaptive immunity.
   - **Tonsils** and **Peyer's patches** in the gastrointestinal tract protect against ingested or inhaled pathogens.

Role in Circulation

One of the lymphatic system's primary roles is to assist in maintaining fluid balance by returning excess interstitial fluid to the bloodstream. Capillary beds in the cardiovascular system continuously exchange water, nutrients, and waste products between blood and tissues. However, about 10% of this fluid remains in the interstitial space, leading to potential swelling if not removed. The lymphatic system absorbs this excess fluid and prevents **edema**, ensuring that tissues do not become overly swollen.

The transport of **large molecules**, such as proteins and lipids, is another critical circulatory function. Plasma proteins that leak from blood capillaries into interstitial fluid are too large to re-enter blood vessels directly. The lymphatic system collects these proteins and returns them to the bloodstream, maintaining osmotic balance and preventing protein loss.

In the digestive system, specialized lymphatic capillaries called **lacteals** absorb dietary fats and fat-soluble vitamins from the small intestine. These fats are transported as **chylomicrons** within the lymph, forming a milky fluid called **chyle**. The lymphatic system delivers these nutrients to the bloodstream via the thoracic duct, integrating them into systemic circulation.

Role in Immunity

The lymphatic system is a central component of the immune system, acting as a surveillance network that detects and responds to pathogens. Lymph nodes, lymphoid organs, and circulating lymphocytes form the backbone of this immune defense.

1. **Lymph Nodes as Filters:** As lymph flows through nodes, immune cells such as **B lymphocytes** and **T lymphocytes** scan for antigens. When foreign particles or pathogens are detected, lymphocytes are activated, triggering an immune response. **Macrophages** in the nodes engulf and digest pathogens, presenting their antigens to lymphocytes to enhance immunity.

2. **T and B Cell Activation:** The thymus and bone marrow are primary lymphoid organs where immune cells mature. T cells, developed in the thymus, are responsible for attacking infected or abnormal cells. B cells, produced in the bone marrow, generate antibodies that neutralize pathogens. Once activated, these lymphocytes migrate to lymph nodes and other secondary lymphoid tissues to respond to infections.

3. **Surveillance and Memory:** The lymphatic system continuously surveys the body for threats. After an infection, memory lymphocytes remain in lymph nodes and other tissues, providing long-term immunity against previously encountered pathogens. This adaptive response ensures faster and stronger defenses upon re-exposure.

Lymphatic System Disorders

Disruptions in the lymphatic system can lead to several health issues. **Lymphedema** occurs when lymphatic vessels fail to drain interstitial fluid effectively, causing swelling in affected areas. This condition can result from injury, infection, or the removal of lymph nodes during cancer treatment.

Infections can also affect the lymphatic system. For example, **lymphangitis** is inflammation of lymphatic vessels caused by bacterial infection, while **filariasis**, a parasitic disease, leads to severe lymphedema (elephantiasis) when parasitic worms block lymphatic vessels.

Cancers of the lymphatic system, such as **Hodgkin's lymphoma** and **non-Hodgkin's lymphoma**, arise from uncontrolled growth of lymphocytes. These conditions can impair immune function and lymphatic circulation, requiring targeted treatments like chemotherapy and radiation.

CHAPTER 11: THE RESPIRATORY SYSTEM

Anatomy of the Respiratory System: From Nose to Alveoli

The respiratory system is designed to bring oxygen into the body and remove carbon dioxide, a waste product of metabolism. It includes a series of structures that filter, warm, and deliver air to the lungs, where gas exchange occurs. Understanding the anatomy of the respiratory system, from the nose to the alveoli, provides insight into how the body maintains oxygen supply and expels carbon dioxide efficiently.

The Upper Respiratory Tract

The upper respiratory tract consists of the nose, nasal cavity, sinuses, pharynx, and larynx. These structures prepare incoming air for the lower respiratory tract.

1. **Nose and Nasal Cavity:** Air enters the respiratory system through the nostrils (**nares**), which open into the nasal cavity. The nasal cavity is lined with **mucous membranes** and tiny hair-like structures called **cilia**. These structures filter dust, pollen, and pathogens from the air. The mucosa also warms and humidifies the air, preventing irritation to the delicate tissues in the lower respiratory tract. The nasal cavity is divided by the **nasal septum** and contains **conchae** (turbinates), bony projections that increase the surface area. This structure ensures efficient conditioning of the air. The nasal cavity connects to the **paranasal sinuses**, air-filled spaces that lighten the skull and produce mucus to trap additional debris.

2. **Pharynx:** Air passes from the nasal cavity into the **pharynx**, a muscular tube shared with the digestive system. The pharynx has three regions:
   - **Nasopharynx:** Located behind the nasal cavity, it serves only as an air passage.
   - **Oropharynx:** Shared by the respiratory and digestive systems, it allows both air and food to pass.
   - **Laryngopharynx:** The lowest section, where air is directed toward the larynx and food toward the esophagus.

3. **Larynx:** The larynx, or voice box, is located at the entrance of the lower respiratory tract. It houses the **vocal cords**, which vibrate to produce sound. The larynx also acts as a protective structure, keeping food and liquids out of the airway during swallowing. The **epiglottis**, a flap of elastic cartilage, covers the opening to the larynx (the **glottis**) when swallowing occurs.

The Lower Respiratory Tract

The lower respiratory tract begins at the trachea and includes the bronchi, bronchioles, and alveoli. These structures form a branching system called the **respiratory tree**, culminating in the alveoli, where gas exchange takes place.

1. **Trachea:** The trachea, or windpipe, is a rigid tube approximately 12 centimeters long and 2.5 centimeters in diameter. It is reinforced by **C-shaped cartilage rings** that keep the airway open during breathing. The trachea is lined with **pseudostratified ciliated columnar epithelium**, which traps particles and moves mucus upward toward the pharynx for expulsion or swallowing. The trachea divides into the **right and left primary bronchi** at the **carina**, a ridge of cartilage that triggers coughing if foreign particles are detected.
2. **Bronchi and Bronchial Tree:** The primary bronchi enter the lungs and branch into **secondary (lobar) bronchi**, each serving one lobe of the lung. The right lung has three lobes, while the left lung has two. Secondary bronchi further divide into **tertiary (segmental) bronchi**, which supply smaller regions within each lobe. Bronchi walls contain cartilage for support and smooth muscle to regulate airflow. As the bronchi branch into smaller **bronchioles**, cartilage diminishes, and smooth muscle becomes more prominent. The bronchioles end in **terminal bronchioles**, the last part of the conducting zone.
3. **Respiratory Bronchioles and Alveolar Ducts:** The respiratory bronchioles mark the beginning of the respiratory zone, where gas exchange occurs. These structures lead into **alveolar ducts**, which are lined with clusters of tiny air sacs called alveoli.

Alveoli: The Site of Gas Exchange

The **alveoli** are microscopic air sacs in the lungs where oxygen and carbon dioxide are exchanged between the air and blood. Each lung contains approximately 300 million alveoli, providing a large surface area for efficient gas exchange.

Alveoli are lined with **simple squamous epithelium**, which allows for easy diffusion of gases. Surrounding the alveoli are **pulmonary capillaries**, where blood from the pulmonary arteries is oxygenated. Oxygen diffuses from the alveoli into the blood, while carbon dioxide moves from the blood into the alveoli to be exhaled.

The alveolar walls also contain **type II alveolar cells**, which secrete **surfactant**, a substance that reduces surface tension and prevents alveolar collapse during exhalation. **Alveolar macrophages** patrol the air sacs, engulfing debris and pathogens to keep the lungs clean.

The Lungs

The **lungs** are paired organs located in the thoracic cavity, flanking the heart. Each lung is encased in a double-layered membrane called the **pleura**, consisting of:

- **Visceral pleura:** The inner layer attached to the lung surface.
- **Parietal pleura:** The outer layer lining the thoracic cavity.

Between these layers is the **pleural cavity**, which contains pleural fluid. This fluid reduces friction during breathing and helps the lungs adhere to the chest wall, ensuring they expand and contract with the movements of the diaphragm and rib cage.

The right lung is larger than the left to accommodate the heart. It is divided into three lobes (superior, middle, and inferior), while the left lung has two lobes (superior and inferior) and a depression called the **cardiac notch** for the heart.

The Diaphragm and Respiratory Muscles

Breathing is powered by the **diaphragm**, a dome-shaped muscle located beneath the lungs. During inhalation, the diaphragm contracts and flattens, increasing the thoracic cavity's volume and drawing air into the lungs. The **external intercostal muscles** assist by lifting the rib cage, further expanding the chest.

During exhalation, the diaphragm and intercostal muscles relax, reducing thoracic volume and pushing air out of the lungs. In forced breathing, additional muscles, such as the **abdominals** and **internal intercostals**, enhance exhalation.

Gas Exchange: Oxygen and Carbon Dioxide Transport

Gas exchange, the primary function of the respiratory system, involves the transfer of oxygen (O_2) into the blood and the removal of carbon dioxide (CO_2) from it. This process occurs at two critical sites: the alveoli in the lungs (external respiration) and the tissues throughout the body (internal respiration). Efficient gas exchange ensures cells receive oxygen for energy production and eliminate CO_2, a byproduct of metabolism.

Gas Exchange at the Alveoli

Oxygen and carbon dioxide move between the alveoli and the pulmonary capillaries through **diffusion**, driven by concentration gradients. The partial pressure of oxygen (PO_2) is higher in the alveoli than in the blood arriving from the pulmonary arteries, causing oxygen to diffuse into the blood. Conversely, the partial pressure of carbon dioxide (PCO_2) is higher in the blood than in the alveoli, driving CO_2 out of the blood and into the alveolar air to be exhaled.

The structure of the alveoli maximizes gas exchange efficiency. The thin walls of **type I alveolar cells**, combined with the single-layered endothelium of the pulmonary capillaries, create a minimal barrier for gas diffusion. The large surface area of approximately 70 square meters in both lungs enhances this exchange.

Surfactant, produced by **type II alveolar cells**, prevents alveolar collapse, maintaining the surface area for effective gas exchange during both inhalation and exhalation. Oxygen-enriched blood from the pulmonary veins then enters the left atrium, ready for systemic circulation.

Oxygen Transport

Once oxygen enters the bloodstream, it is transported primarily by **hemoglobin**, a protein in red blood cells. Each hemoglobin molecule contains four iron-containing **heme groups**, each capable of binding one oxygen molecule. This reversible binding allows oxygen to be released where it is needed most.

The oxygen-hemoglobin dissociation curve illustrates how hemoglobin's affinity for oxygen changes with varying PO_2 levels. In the lungs, where PO_2 is high, hemoglobin binds oxygen efficiently. In tissues with lower PO_2, such as active muscles, hemoglobin releases oxygen to meet cellular demands.

A small amount of oxygen (about 1.5%) is dissolved directly in the plasma, which contributes to the blood's PO_2 and facilitates diffusion into tissues.

Carbon Dioxide Transport

Carbon dioxide, produced as a waste product of cellular respiration, is transported from tissues to the lungs in three forms:

1. **Bicarbonate Ions (HCO_3^-):** The majority of CO_2 (about 70%) is converted to bicarbonate in red blood cells. The enzyme **carbonic anhydrase** catalyzes the reaction: $CO_2 + H_2O \rightleftharpoons H_2CO_3 \rightleftharpoons H^+ + HCO_3^-$. Bicarbonate ions diffuse into the plasma, acting as a buffer to regulate blood pH.
2. **Carbaminohemoglobin:** About 20-23% of CO_2 binds to hemoglobin's protein portion (not the heme groups), forming carbaminohemoglobin. This binding is enhanced in tissues where oxygen has been released.
3. **Dissolved CO_2:** The remaining 7-10% of CO_2 is transported dissolved in plasma.

In the lungs, these processes reverse. Bicarbonate ions re-enter red blood cells, where carbonic anhydrase facilitates the conversion of HCO_3^- back into CO_2. This CO_2, along with CO_2 released from carbaminohemoglobin and plasma, diffuses into the alveoli for exhalation.

Internal Respiration

At the tissue level, oxygen diffuses from systemic capillaries into cells, driven by a lower PO_2 in the interstitial fluid and mitochondria. Simultaneously, CO_2 produced by cellular metabolism diffuses into the blood, maintaining the exchange gradient for continuous gas transport.

The Role of the Respiratory System in Acid-Base Balance

The respiratory system is critical in maintaining **acid-base balance**, a vital aspect of homeostasis. Blood pH must remain within a narrow range (7.35–7.45) for enzymes and metabolic processes to function properly. The respiratory system achieves this by regulating the concentration of carbon dioxide, which directly affects the levels of carbonic acid and bicarbonate in the blood.

The Carbonic Acid-Bicarbonate Buffer System

The primary mechanism for blood pH regulation is the **carbonic acid-bicarbonate buffer system**, which relies on the reversible reaction:

$$CO_2 + H_2O \rightleftharpoons H_2CO_3 \rightleftharpoons H^+ + HCO_3^-$$

Carbon dioxide reacts with water in red blood cells to form carbonic acid (H_2CO_3), which dissociates into hydrogen ions (H^+) and bicarbonate ions (HCO_3^-). An increase in CO_2 levels raises H^+ concentration, lowering blood pH (making it more acidic). Conversely, a decrease in CO_2 reduces H^+ concentration, raising pH (making it more alkaline).

The respiratory system adjusts CO_2 levels by modifying the rate and depth of breathing:

- **Hyperventilation:** Rapid, deep breathing expels more CO_2, reducing its concentration in the blood. This shift lowers H^+ levels, increasing pH (respiratory alkalosis).
- **Hypoventilation:** Slow, shallow breathing retains CO_2, increasing its concentration in the blood. This raises H^+ levels, lowering pH (respiratory acidosis).

Regulation by Chemoreceptors

Chemoreceptors in the body monitor blood pH and CO_2 levels to regulate respiratory activity:

- **Central chemoreceptors** in the medulla oblongata respond to changes in CO_2 and pH in cerebrospinal fluid. Even slight increases in CO_2 stimulate these receptors to increase ventilation.
- **Peripheral chemoreceptors** in the carotid and aortic bodies detect changes in blood pH, PO_2, and PCO_2. While primarily sensitive to CO_2, they also increase respiratory rate in response to significant drops in oxygen levels.

These feedback mechanisms ensure that blood pH remains stable, even during physical exertion or metabolic stress.

Respiratory Compensation

The respiratory system compensates for metabolic disturbances to maintain acid-base balance. For example, during **metabolic acidosis**, such as in uncontrolled diabetes (ketoacidosis), excess H^+ lowers blood pH. The respiratory system responds by increasing ventilation to eliminate CO_2, reducing H^+ levels and partially correcting the acidosis.

Similarly, during **metabolic alkalosis**, caused by excessive vomiting or bicarbonate intake, the respiratory system slows breathing to retain CO_2 and restore pH balance. However, respiratory compensation is limited; prolonged imbalances may require renal adjustments.

Disorders Affecting Acid-Base Balance

Respiratory diseases can disrupt acid-base regulation:

- **Chronic obstructive pulmonary disease (COPD):** Impaired gas exchange leads to CO_2 retention, causing chronic respiratory acidosis. Kidneys may compensate by increasing bicarbonate retention, but the imbalance often persists.
- **Acute respiratory distress syndrome (ARDS):** Severe inflammation or damage to the lungs reduces oxygenation and CO_2 elimination, exacerbating acidosis.
- **Hyperventilation syndrome:** Excessive breathing reduces CO_2 levels, causing respiratory alkalosis. Symptoms include dizziness, tingling, and muscle spasms.

In all cases, acid-base disturbances require careful management to restore normal respiratory and metabolic function.

CHAPTER 12: THE DIGESTIVE SYSTEM

The Digestive Tract: Anatomy and Functions of Each Organ

The digestive tract is a continuous tube that processes food, extracts nutrients, and eliminates waste. It starts at the mouth and ends at the anus, with each organ in between contributing specific functions to break down food and absorb nutrients. Understanding the anatomy and role of each part highlights the efficiency of this system in sustaining the body.

Mouth

The digestive process begins in the **mouth**, where food is ingested. The **teeth** break food into smaller pieces through mechanical digestion, increasing the surface area for enzymes to act. The **tongue** helps mix food with saliva, forming a soft mass called a **bolus** that is easier to swallow.

Saliva, produced by the **salivary glands**, contains **amylase**, an enzyme that starts breaking down carbohydrates into simpler sugars. Saliva also moistens food, making it easier to move through the esophagus. The **hard palate** forms the roof of the mouth, while the **soft palate** and **uvula** prevent food from entering the nasal cavity during swallowing.

Pharynx and Esophagus

After swallowing, the bolus passes into the **pharynx**, a shared pathway for food and air. The **epiglottis**, a flap of tissue, prevents food from entering the trachea by covering its opening. The bolus then moves into the **esophagus**, a muscular tube that connects the pharynx to the stomach.

The esophagus transports food using **peristalsis**, a series of coordinated muscle contractions that push the bolus downward. Its walls are lined with smooth muscle and mucosa, ensuring efficient movement and protecting the tissue from abrasion. At the end of the esophagus is the **lower esophageal sphincter (LES)**, a ring of muscle that prevents stomach acid from flowing back into the esophagus.

Stomach

The **stomach** is a muscular, J-shaped organ located in the upper abdomen. It serves as a temporary storage site and the primary location for protein digestion. The stomach's interior is lined with **rugae**, folds that allow it to expand as it fills with food.

Gastric glands in the stomach lining secrete **gastric juice**, a combination of hydrochloric acid (HCl), pepsinogen, and mucus. HCl creates an acidic environment (pH 1.5–3.5) that denatures proteins and activates **pepsin**, an enzyme that breaks proteins into smaller peptides. The mucus protects the stomach lining from being digested by its own enzymes.

The stomach's muscular walls mix food with gastric juice, forming a semi-liquid substance called **chyme**. The **pyloric sphincter**, located at the stomach's lower end, regulates the release of chyme into the small intestine.

Small Intestine

The **small intestine** is the longest part of the digestive tract, measuring about 6 meters in adults. It is divided into three sections: the **duodenum, jejunum**, and **ileum**. The small intestine is the primary site for nutrient digestion and absorption.

1. **Duodenum:** This first section receives chyme from the stomach, along with bile from the liver and pancreatic juice from the pancreas. **Bile**, stored in the gallbladder, emulsifies fats, breaking them into smaller droplets for easier digestion. Pancreatic juice contains enzymes like **lipase**, **amylase**, and **proteases**, which digest fats, carbohydrates, and proteins, respectively. The duodenum also neutralizes stomach acid with bicarbonate, protecting the intestinal lining.

2. **Jejunum:** Most nutrient absorption occurs in the jejunum. Its inner walls are covered with **villi**, finger-like projections that increase the surface area for absorption. Each villus contains blood capillaries and a lacteal, which transport absorbed nutrients to the bloodstream and lymphatic system.

3. **Ileum:** The final section continues absorbing nutrients, including vitamin B12 and bile salts. The ileum ends at the **ileocecal valve**, which controls the passage of chyme into the large intestine.

Large Intestine

The **large intestine** is shorter than the small intestine but wider in diameter. It consists of the **cecum, colon, rectum**, and **anal canal**. Its primary functions are to absorb water and electrolytes, form feces, and house beneficial bacteria.

1. **Cecum:** The cecum is a pouch-like structure that connects the ileum to the colon. The **appendix**, a small, finger-shaped structure attached to the cecum, is thought to have a role in maintaining gut flora, although it is not essential for digestion.

2. **Colon:** The colon is divided into ascending, transverse, descending, and sigmoid sections. As chyme moves through the colon, water and electrolytes are absorbed, converting it into solid feces. The colon is home

to trillions of bacteria that ferment undigested carbohydrates, producing short-chain fatty acids and vitamins like **vitamin K**.

3. **Rectum and Anal Canal:** The rectum stores feces until defecation. The anal canal has two sphincters, one involuntary (internal) and one voluntary (external), that regulate the release of feces. This ensures control over bowel movements.

Accessory Organs

Although not part of the digestive tract, **accessory organs** like the liver, pancreas, and gallbladder contribute significantly to digestion.

1. **Liver:** The liver produces bile, processes absorbed nutrients, and detoxifies harmful substances. Blood from the digestive tract flows through the **hepatic portal vein** to the liver, allowing it to metabolize glucose, lipids, and proteins.

2. **Gallbladder:** The gallbladder stores and concentrates bile, releasing it into the duodenum when fats are present. Bile emulsifies fats, making them easier to digest and absorb.

3. **Pancreas:** The pancreas secretes digestive enzymes and bicarbonate into the duodenum. It also produces insulin and glucagon, hormones that regulate blood sugar levels.

The digestive tract's intricate anatomy ensures that food is broken down efficiently, nutrients are absorbed, and waste is expelled. Each organ, from the mouth to the anus, contributes a specific function, working together to nourish the body and maintain overall health.

Enzymes and Nutrient Absorption

Enzymes are essential for breaking down complex food molecules into smaller, absorbable units. Each enzyme is specific to a type of nutrient, such as carbohydrates, proteins, or fats. These enzymes work in various parts of the digestive tract, ensuring efficient digestion and absorption.

Carbohydrate Digestion and Absorption

Carbohydrate digestion begins in the **mouth**, where **salivary amylase** breaks down starch into maltose and smaller polysaccharides. As food enters the stomach, the acidic environment inactivates salivary amylase, halting carbohydrate digestion temporarily. The process resumes in the **small intestine**, where **pancreatic**

amylase, secreted by the pancreas, continues breaking down polysaccharides into disaccharides like maltose, sucrose, and lactose.

The final stage occurs at the surface of the **intestinal brush border**, where enzymes like **maltase**, **sucrase**, and **lactase** break disaccharides into monosaccharides such as glucose, fructose, and galactose. These monosaccharides are absorbed through the walls of the small intestine into the bloodstream via **active transport** and **facilitated diffusion**. Glucose and galactose are actively transported using sodium-dependent glucose transporters, while fructose enters cells via facilitated diffusion through GLUT transporters.

Protein Digestion and Absorption

Protein digestion begins in the **stomach**, where **pepsin**, activated by hydrochloric acid, breaks proteins into smaller peptides. Pepsin works optimally in the stomach's acidic environment but becomes inactive when chyme enters the alkaline **duodenum**.

In the small intestine, pancreatic enzymes **trypsin**, **chymotrypsin**, and **carboxypeptidase** further break peptides into shorter chains and free amino acids. The pancreas secretes these enzymes in inactive forms, activated in the duodenum by **enterokinase**, an enzyme produced by the intestinal lining. Brush-border enzymes like **aminopeptidase** and **dipeptidase** complete protein digestion, releasing individual amino acids.

Amino acids are absorbed into the bloodstream through **active transport mechanisms**. Sodium and hydrogen ion gradients drive amino acid transport across the intestinal epithelium. Once inside the bloodstream, amino acids are delivered to cells for protein synthesis or energy production.

Fat Digestion and Absorption

Fat digestion requires bile and enzymes to overcome the insolubility of lipids in water. In the **duodenum**, bile salts from the liver emulsify fats into tiny droplets called **micelles**, increasing the surface area for enzymes to act. **Pancreatic lipase**, secreted into the small intestine, breaks triglycerides into monoglycerides and free fatty acids.

These products are absorbed through the intestinal epithelial cells by diffusion. Inside the cells, monoglycerides and fatty acids are reassembled into triglycerides and packaged into **chylomicrons**, which are transported via the **lymphatic system**. Chylomicrons enter the bloodstream through the thoracic duct, delivering fats to tissues for energy storage or use.

Nucleic Acid Digestion and Absorption

Nucleic acids, such as DNA and RNA, are digested by **deoxyribonuclease (DNase)** and **ribonuclease (RNase)**, enzymes secreted by the pancreas. These enzymes break nucleic acids into nucleotides. Brush-border enzymes like **nucleotidase** and **nucleosidase** further degrade nucleotides into nitrogenous bases, sugars, and phosphate groups, which are absorbed into the bloodstream.

The Role of Gut Flora in Digestion

The gut flora, or **gut microbiota**, consists of trillions of microorganisms, including bacteria, viruses, and fungi, that reside in the digestive tract. These microbes contribute to digestion, immune function, and overall health. Most gut flora is found in the **large intestine**, where they ferment undigested food and produce beneficial compounds.

Fermentation of Dietary Fiber

Humans lack the enzymes to digest certain carbohydrates, such as cellulose and resistant starch, found in plant-based foods. Gut bacteria ferment these fibers in the colon, producing **short-chain fatty acids (SCFAs)** like acetate, propionate, and butyrate. SCFAs provide an energy source for colon cells, reduce inflammation, and regulate gut motility.

Butyrate is particularly important for maintaining the integrity of the **intestinal barrier**. It strengthens tight junctions between epithelial cells, preventing harmful substances from entering the bloodstream. SCFAs also modulate immune responses and influence metabolic processes in distant organs.

Production of Vitamins

Gut bacteria synthesize several vitamins essential for health. **Vitamin K**, produced by species like *Escherichia coli*, is necessary for blood clotting. Some gut bacteria also produce **B vitamins**, including biotin (B7) and folate (B9), which are absorbed in the large intestine. These vitamins support energy metabolism, DNA synthesis, and cellular function.

Protein Metabolism by Gut Flora

Gut bacteria degrade undigested proteins and amino acids, producing compounds such as **ammonia**, **indoles**, and **hydrogen sulfide**. While excessive production of these byproducts can be harmful, gut microbes also help recycle nitrogen and produce metabolites that influence host metabolism.

Immune System Modulation

The gut microbiota interacts closely with the immune system, training it to distinguish between harmful pathogens and benign or beneficial microbes. Bacteria like *Bacteroides fragilis* stimulate the production of **regulatory T cells**, which prevent overactive immune responses. The microbiota also supports the production of **mucus** and **antimicrobial peptides**, which protect the gut lining from infections.

Competition with Pathogens

Gut flora helps maintain a healthy balance of microbes by competing with harmful bacteria for nutrients and attachment sites. Beneficial bacteria produce **bacteriocins**, antimicrobial substances that inhibit the growth of pathogens. A healthy microbiota can prevent infections caused by bacteria like *Clostridioides difficile*, which thrive when the gut's microbial balance is disrupted.

Dysbiosis and Health Implications

An imbalance in gut flora, or **dysbiosis**, can lead to digestive disorders and systemic diseases. For example, reduced microbial diversity is associated with conditions like **irritable bowel syndrome (IBS)**, **inflammatory bowel disease (IBD)**, and obesity. Dysbiosis may result from antibiotic use, poor diet, or chronic stress, all of which disrupt microbial communities.

Restoring gut health often involves dietary interventions, such as increasing fiber intake or consuming **probiotics** and **prebiotics**. Probiotics, found in fermented foods like yogurt and kefir, introduce beneficial bacteria, while prebiotics, such as inulin and fructooligosaccharides, provide nutrients that promote the growth of beneficial microbes.

The gut flora forms a dynamic partnership with the digestive system, contributing to nutrient extraction, immune defense, and overall health. Its balance and diversity are essential for optimal digestive function and systemic well-being.

CHAPTER 13: THE URINARY SYSTEM

Anatomy and Function of the Kidneys

The **kidneys** are bean-shaped organs located on either side of the spine, just below the rib cage. Each kidney is about the size of a fist and lies in the retroperitoneal space, partially protected by the lower ribs. The kidneys filter blood, remove waste products, regulate fluid balance, and maintain electrolyte and acid-base homeostasis. Understanding their anatomy and function highlights their importance in sustaining life.

Gross Anatomy of the Kidneys

The kidney has three main regions: the **cortex, medulla,** and **renal pelvis**.

- The **renal cortex** is the outer layer, where filtration begins. It contains the **glomeruli** and parts of the nephrons, the functional units of the kidney.
- The **renal medulla** lies beneath the cortex and consists of pyramid-shaped structures called **renal pyramids**. These pyramids contain the loops of Henle and collecting ducts, which concentrate urine.
- The **renal pelvis** is a funnel-shaped cavity that collects urine from the medulla and channels it into the **ureter**, which transports urine to the bladder.

The outer surface of the kidney is covered by a tough, fibrous capsule that provides protection. Surrounding this capsule is a layer of fat that cushions the kidney and helps anchor it in place.

Blood Supply to the Kidneys

The kidneys receive about 20-25% of the body's cardiac output, reflecting their high demand for blood. Blood enters each kidney through the **renal artery**, a branch of the abdominal aorta. Inside the kidney, the renal artery branches into smaller **afferent arterioles**, which supply blood to the **glomeruli**. After filtration, blood leaves the glomeruli through **efferent arterioles** and enters the **peritubular capillaries** and **vasa recta**, networks that surround the nephron and aid in reabsorption and secretion.

Filtered blood exits the kidney via the **renal vein**, which drains into the inferior vena cava.

The Nephron: The Functional Unit

Each kidney contains approximately 1 million **nephrons**, the microscopic structures responsible for filtering blood and forming urine. A nephron has two main parts: the **renal corpuscle** and the **renal tubule**.

1. **Renal Corpuscle:**

    - The renal corpuscle consists of the **glomerulus**, a network of capillaries, and the **Bowman's capsule**, a cup-shaped structure that surrounds the glomerulus. Blood pressure forces water, ions, glucose, and waste products from the glomerulus into Bowman's capsule, forming a filtrate.
    - Large molecules like proteins and blood cells are retained in the bloodstream, ensuring only small solutes enter the filtrate.

2. **Renal Tubule:**

    - The filtrate flows into the renal tubule, which has three segments: the **proximal convoluted tubule (PCT)**, the **loop of Henle**, and the **distal convoluted tubule (DCT)**.
    - The **PCT** reabsorbs about 65-70% of the filtrate, including glucose, amino acids, sodium, and water. It also secretes substances like hydrogen ions and drugs into the tubule.
    - The **loop of Henle** concentrates the filtrate by creating a gradient that allows water to leave the descending limb and salts to leave the ascending limb.
    - The **DCT** fine-tunes ion reabsorption under the influence of hormones like **aldosterone** and **parathyroid hormone (PTH)**.

The tubule ends in a **collecting duct**, which collects filtrate from multiple nephrons. The collecting duct further concentrates urine under the influence of **antidiuretic hormone (ADH)**, which increases water reabsorption.

Functions of the Kidneys

1. **Filtration of Blood:** The kidneys filter about 50 gallons of blood per day, removing waste products like **urea**, **creatinine**, and toxins. This filtration occurs in the glomeruli, where hydrostatic pressure drives substances into Bowman's capsule. The kidneys ensure that essential substances like glucose and amino acids are reabsorbed while waste is excreted.

2. **Regulation of Fluid Balance:** The kidneys control fluid levels by adjusting water reabsorption. When the body is dehydrated, ADH increases, promoting water reabsorption in the collecting ducts. When hydration levels are adequate, less ADH is released, allowing excess water to be excreted as dilute urine.

3. **Electrolyte Balance:** The kidneys regulate key electrolytes like sodium, potassium, calcium, and chloride. Aldosterone, released by the adrenal glands, stimulates the reabsorption of sodium and the excretion of potassium in the DCT. This balance is essential for nerve function, muscle contraction, and maintaining blood pressure.

4. **Acid-Base Regulation:** The kidneys maintain pH balance by reabsorbing bicarbonate (HCO_3^-) and excreting hydrogen ions (H^+). When blood pH drops (acidosis), the kidneys increase bicarbonate reabsorption and hydrogen ion secretion. During alkalosis, bicarbonate excretion increases.

5. **Hormone Production:** The kidneys produce hormones that influence other systems:
    - **Erythropoietin (EPO):** Stimulates red blood cell production in the bone marrow in response to low oxygen levels.
    - **Renin:** Released by the juxtaglomerular apparatus in response to low blood pressure, renin activates the **renin-angiotensin-aldosterone system (RAAS)**, increasing blood pressure and sodium retention.
    - **Calcitriol (active vitamin D):** Promotes calcium absorption in the intestines and supports bone health.

6. **Excretion of Waste Products:** Urea, the main nitrogenous waste, is formed in the liver and excreted by the kidneys. Other wastes, like creatinine from muscle metabolism and uric acid from nucleic acid breakdown, are also eliminated in urine.

How the Body Maintains Water and Electrolyte Balance

The body maintains water and electrolyte balance through the coordinated actions of the kidneys, hormones, and feedback mechanisms. These processes ensure that the body retains the right amount of fluids and maintains optimal concentrations of sodium, potassium, calcium, and other ions. Proper balance is essential for nerve function, muscle contraction, and overall cellular activity.

Water Balance

Water balance is achieved by matching water intake with water loss. The kidneys are the primary regulators of water in the body, adjusting urine output based on hydration status. When the body is dehydrated, the **hypothalamus** detects an increase in blood osmolality, stimulating the release of **antidiuretic hormone (ADH)** from the posterior pituitary. ADH acts on the **collecting ducts** of the kidneys, making them more permeable to water. This allows water to be reabsorbed into the bloodstream, producing concentrated urine and conserving body water.

Conversely, in overhydration, ADH secretion decreases, reducing water reabsorption. The collecting ducts allow excess water to be excreted, resulting in dilute urine. This precise control ensures that blood osmolality remains within a narrow range, protecting cells from swelling or shrinking.

Electrolyte Balance

Sodium, potassium, and calcium are the key electrolytes regulated by the kidneys.

- **Sodium (Na^+):** Sodium is the primary extracellular ion and is critical for fluid balance and nerve transmission. The hormone **aldosterone**, produced by the adrenal glands, stimulates sodium reabsorption in the **distal convoluted tubule (DCT)** and **collecting ducts**. When blood sodium levels are low or blood pressure drops, the **renin-angiotensin-aldosterone system (RAAS)** is activated. This system increases aldosterone secretion, promoting sodium retention and water reabsorption, which raises blood pressure and restores balance.

- **Potassium (K^+):** Potassium, the main intracellular ion, is crucial for muscle contraction and nerve signaling. The kidneys excrete excess potassium to prevent hyperkalemia, which can cause arrhythmias. Aldosterone also regulates potassium levels by promoting its excretion in the DCT and collecting ducts.

- **Calcium (Ca^{2+}):** Calcium levels are regulated by **parathyroid hormone (PTH)**, which increases calcium reabsorption in the kidneys, promotes calcium release from bones, and enhances calcium absorption in the intestines by activating vitamin D. This coordination ensures calcium levels remain stable for bone health and cellular processes.

Fluid and Electrolyte Imbalances

Disruptions in water or electrolyte balance can have serious consequences. **Dehydration** occurs when water loss exceeds intake, leading to low blood volume and high blood osmolality. Symptoms include dry mouth, fatigue, and decreased urine output. Severe dehydration can cause kidney damage or organ failure.

Hyponatremia, or low sodium levels, can result from excessive water intake or loss of sodium through sweating or diarrhea. It disrupts nerve and muscle function, leading to confusion, seizures, or even coma. **Hypernatremia**, or high sodium levels, causes cellular dehydration, triggering thirst, and neurological symptoms.

Maintaining water and electrolyte balance depends on the kidneys' ability to adjust filtration, reabsorption, and excretion in response to the body's needs. Hormonal regulation and feedback loops ensure homeostasis under varying conditions.

The Process of Urine Formation

Urine formation occurs in the nephrons of the kidneys and involves three key processes: **filtration, reabsorption**, and **secretion**. These processes ensure that the body eliminates waste products while conserving essential nutrients and maintaining fluid and electrolyte balance.

Filtration

Filtration begins in the **glomerulus**, a network of capillaries located within the renal corpuscle. Blood enters the glomerulus through the **afferent arteriole** and exits via the **efferent arteriole**. The high hydrostatic pressure in the glomerulus forces water, ions, glucose, amino acids, and waste products into **Bowman's capsule**, forming the filtrate.

The filtration barrier, consisting of the glomerular endothelium, basement membrane, and podocytes, prevents large molecules like proteins and blood cells from entering the filtrate. Approximately 180 liters of filtrate are produced daily, but most of it is reabsorbed during subsequent processes.

Reabsorption

Reabsorption occurs primarily in the **proximal convoluted tubule (PCT)**, where about 65% of the filtrate is reclaimed. Nutrients like glucose, amino acids, and ions are actively transported back into the bloodstream. Water follows these solutes via osmosis, ensuring that the body retains essential components.

In the **loop of Henle**, reabsorption continues. The descending limb is permeable to water but not solutes, allowing water to exit into the medulla's hypertonic environment. The ascending limb is impermeable to water but actively transports sodium, potassium, and chloride into the medulla. This countercurrent mechanism concentrates the filtrate and establishes a gradient for further water reabsorption.

In the **distal convoluted tubule (DCT)** and **collecting duct**, reabsorption is fine-tuned by hormones like ADH and aldosterone. These structures adjust the amount of water and electrolytes reabsorbed based on the body's needs.

Secretion

Secretion involves the active transport of additional substances from the blood into the filtrate. In the PCT, substances like hydrogen ions, ammonia, and certain drugs are secreted to maintain pH balance and eliminate toxins. The DCT and collecting ducts secrete potassium and hydrogen ions under hormonal control, helping regulate electrolyte and acid-base balance.

Final Urine Composition

By the time filtrate reaches the end of the collecting duct, it has become urine. Urine consists of water (95%), urea, creatinine, uric acid, and trace amounts of electrolytes and other waste products. It flows into the **renal pelvis**, down the **ureters**, and into the **bladder** for storage before excretion.

Acid-Base Regulation in the Kidneys

The kidneys regulate acid-base balance by controlling the concentration of hydrogen ions (H^+) and bicarbonate ions (HCO_3^-) in the blood. This balance is essential for maintaining a pH range of 7.35–7.45, which is critical for enzymatic activity and cellular function.

Buffer Systems and pH Maintenance

The primary mechanism for pH regulation is the **carbonic acid-bicarbonate buffer system**:

$$CO_2 + H_2O \rightleftharpoons H_2CO_3 \rightleftharpoons H^+ + HCO_3^-$$

When blood pH drops (acidosis), the kidneys excrete more hydrogen ions and reabsorb bicarbonate, neutralizing excess acidity. In alkalosis, the kidneys retain hydrogen ions and excrete bicarbonate, lowering pH to normal levels.

Hydrogen Ion Secretion

Hydrogen ions are secreted into the renal tubules, primarily in the **proximal convoluted tubule (PCT)** and **distal convoluted tubule (DCT)**. In the PCT, hydrogen ions combine with bicarbonate in the filtrate to form carbonic acid, which dissociates into water and carbon dioxide. These molecules diffuse back into the tubular cells, where they recombine to form bicarbonate and hydrogen ions. The bicarbonate is reabsorbed into the blood, while hydrogen ions are secreted back into the filtrate.

In the DCT and collecting ducts, hydrogen ions are actively secreted by **intercalated cells**, which use ATP-driven pumps. These secreted hydrogen ions bind to buffers like ammonia or phosphate in the filtrate, preventing excessive acidification of urine.

Bicarbonate Reabsorption

The kidneys reabsorb nearly all the filtered bicarbonate to maintain acid-base balance. This process occurs in the PCT, where bicarbonate ions are converted into carbonic acid in the filtrate. After dissociation into water and carbon dioxide, these components are reabsorbed and used to regenerate bicarbonate inside tubular cells. The regenerated bicarbonate is then transported back into the bloodstream.

Ammonia Production and Excretion

When the body experiences prolonged acidosis, the kidneys increase the production of **ammonia** from glutamine in the renal tubules. Ammonia binds to hydrogen ions in the filtrate to form **ammonium ions (NH_4^+)**, which are excreted in urine. This mechanism allows the kidneys to secrete large amounts of acid without significantly lowering urinary pH.

Clinical Implications

Impaired acid-base regulation can lead to disorders like **metabolic acidosis** or **alkalosis**. Chronic kidney disease reduces the ability to excrete hydrogen ions and reabsorb bicarbonate, causing acidosis. Conversely, conditions like excessive vomiting can result in alkalosis due to bicarbonate retention.

The kidneys' ability to regulate acid-base balance ensures that blood pH remains stable, even under metabolic stress. Their integration with respiratory and buffering systems allows precise control of pH, maintaining the body's internal equilibrium.

CHAPTER 14: THE IMMUNE SYSTEM

Innate vs. Adaptive Immunity: First and Second Lines of Defense

The immune system defends the body against pathogens, toxins, and abnormal cells using two main strategies: **innate immunity** and **adaptive immunity**. Innate immunity provides immediate, non-specific protection, while adaptive immunity offers a delayed, highly specific response that includes memory for future encounters with the same pathogens. These systems work together to protect the body effectively.

Innate Immunity: The First Line of Defense

Innate immunity is the body's initial barrier to infection. It includes physical, chemical, and cellular defenses that respond quickly and broadly to pathogens. Innate immunity does not require prior exposure to a pathogen and does not adapt or improve with repeated exposure.

Physical Barriers: The first layer of defense includes the **skin** and **mucous membranes**, which act as physical shields. The skin's tightly packed epithelial cells form a barrier that prevents pathogens from entering the body. The **keratin** in the outer layer of the skin provides additional toughness, while **sebaceous glands** secrete sebum, a substance that inhibits microbial growth.

Mucous membranes line the respiratory, gastrointestinal, and urogenital tracts. These membranes produce **mucus**, which traps pathogens and prevents them from reaching underlying tissues. In the respiratory tract, **cilia** sweep mucus and trapped particles out of the airways, a process known as the **mucociliary escalator**.

Chemical Barriers: Chemical defenses enhance the effectiveness of physical barriers. **Lysozyme**, an enzyme found in tears, saliva, and sweat, breaks down bacterial cell walls. **Stomach acid** (hydrochloric acid) kills most ingested pathogens. In the skin and mucosa, **antimicrobial peptides**, like defensins, disrupt microbial membranes, further reducing the chance of infection.

Microbial Barriers: The body's natural **microbiota** competes with pathogens for nutrients and space, limiting their ability to establish infections. Beneficial bacteria in the gut and skin produce substances that inhibit pathogenic growth.

Innate Immunity: The Second Line of Defense

If pathogens breach the physical and chemical barriers, the second line of defense comes into play. This includes cellular responses, inflammatory reactions, and the activation of soluble mediators.

Phagocytes: Phagocytic cells, such as **macrophages** and **neutrophils**, engulf and destroy pathogens. These cells recognize pathogens using **pattern recognition receptors (PRRs)**, such as **toll-like receptors (TLRs)**, which detect conserved microbial patterns like bacterial lipopolysaccharides or viral RNA. After engulfing a pathogen, phagocytes break it down with enzymes and present its antigens to adaptive immune cells, bridging innate and adaptive immunity.

Natural Killer (NK) Cells: NK cells target virus-infected cells and tumor cells. Unlike phagocytes, NK cells induce apoptosis (programmed cell death) in abnormal cells. They release perforins, which create pores in the target cell's membrane, and granzymes, which trigger cell death from within.

Inflammatory Response: The **inflammatory response** is a hallmark of innate immunity. When tissues are damaged or infected, they release chemical signals like **histamine** and **prostaglandins**. These signals cause blood vessels to dilate and become more permeable, allowing immune cells to reach the site of infection. This leads to redness, heat, swelling, and pain—signs of inflammation.

Complement System: The **complement system** consists of proteins in the blood that enhance immune responses. When activated, complement proteins form the **membrane attack complex (MAC)**, which punches holes in bacterial membranes, causing lysis. Complement proteins also mark pathogens for destruction (opsonization) and attract immune cells to the site of infection (chemotaxis).

Interferons: Interferons are signaling proteins produced by virus-infected cells. They warn neighboring cells to increase their antiviral defenses and activate immune cells like NK cells. Interferons slow viral replication, buying time for the adaptive immune response to develop.

Adaptive Immunity: The Third Line of Defense

Adaptive immunity provides a specific response tailored to particular pathogens. Unlike innate immunity, it requires time to develop after initial exposure but offers long-lasting protection through **immunological memory**. Adaptive immunity relies on **lymphocytes**, specifically B cells and T cells.

B Cells and Antibody Production: B cells, a type of lymphocyte, recognize antigens using membrane-bound antibodies. Upon activation, they differentiate into **plasma cells**, which secrete large quantities of specific antibodies into the bloodstream. Antibodies perform several functions:

- **Neutralization:** They bind to toxins or viruses, preventing them from interacting with host cells.

- **Opsonization:** They coat pathogens, making them easier for phagocytes to recognize and engulf.
- **Complement Activation:** Antibodies activate the complement system, enhancing pathogen destruction.

Some B cells become **memory B cells**, which remain in the body for years, ready to produce antibodies quickly during subsequent infections by the same pathogen.

T Cells and Cellular Immunity: T cells mediate the cellular arm of adaptive immunity. They mature in the thymus and are divided into two main types:

- **Helper T Cells (CD4$^+$):** These cells coordinate immune responses by releasing cytokines that activate other immune cells, including B cells, cytotoxic T cells, and macrophages.
- **Cytotoxic T Cells (CD8$^+$):** These cells target and kill infected or abnormal cells by recognizing antigens presented on **major histocompatibility complex (MHC)** molecules. Like NK cells, cytotoxic T cells release perforins and granzymes to induce apoptosis.

Antigen Presentation: Adaptive immunity depends on **antigen-presenting cells (APCs)**, such as dendritic cells and macrophages. APCs engulf pathogens, process their antigens, and present them on their surface using MHC molecules. This presentation activates T cells, initiating the adaptive response.

Memory and Vaccination: One of the defining features of adaptive immunity is memory. After an infection, memory B and T cells persist in the body, enabling a faster and stronger response to future infections by the same pathogen. Vaccines exploit this property by introducing harmless antigens to stimulate memory cell formation, providing immunity without causing disease.

Innate and Adaptive Immunity Working Together

Innate and adaptive immunity are deeply interconnected. Innate responses provide immediate protection and guide the development of adaptive immunity. For example, dendritic cells from the innate system present antigens to T cells, activating the adaptive response. Cytokines released by helper T cells enhance innate functions, such as macrophage activation and inflammation.

By combining broad, rapid defenses with precise, long-term strategies, the immune system ensures robust protection against a wide range of threats. Each component, from physical barriers to memory cells, contributes to the body's ability to defend itself effectively.

How Vaccines Work and Their Impact on Health

Vaccines protect the body by training the immune system to recognize and combat harmful pathogens. They introduce antigens—harmless forms or components of a pathogen—into the body, triggering an immune response without causing disease. This process creates **immunological memory**, enabling the body to respond quickly and effectively to future exposures.

Types of Vaccines

Vaccines are designed to mimic natural infections, using different approaches to stimulate immunity:

1. **Live-Attenuated Vaccines:** These vaccines use a weakened form of the pathogen that cannot cause serious illness in healthy individuals. Examples include the measles, mumps, and rubella (MMR) vaccine and the chickenpox vaccine. Live-attenuated vaccines produce a strong, long-lasting immune response because they closely mimic natural infections. However, they are unsuitable for individuals with weakened immune systems.

2. **Inactivated Vaccines:** Inactivated vaccines contain pathogens that have been killed or inactivated. The polio and hepatitis A vaccines are examples. While safer than live vaccines for immunocompromised individuals, inactivated vaccines often require booster shots to maintain immunity because their response is not as robust.

3. **Subunit, Recombinant, and Conjugate Vaccines:** These vaccines use specific pieces of the pathogen, such as proteins or sugars, to trigger an immune response. For example, the human papillomavirus (HPV) vaccine uses viral proteins, while the Haemophilus influenzae type B (Hib) vaccine uses polysaccharides linked to a carrier protein. These vaccines are highly targeted and reduce the risk of side effects.

4. **Messenger RNA (mRNA) Vaccines:** mRNA vaccines, like those developed for COVID-19 (Pfizer-BioNTech and Moderna), deliver genetic instructions for cells to produce a harmless piece of the pathogen, such as the spike protein of SARS-CoV-2. This stimulates an immune response and the production of antibodies. mRNA vaccines are fast to develop and do not contain live virus, making them safe for most people.

5. **Viral Vector Vaccines:** These use a harmless virus to deliver genetic material from the pathogen into the body. The Johnson & Johnson COVID-19 vaccine is an example. The vector does not replicate in the body but serves as a delivery system for triggering immunity.

How Vaccines Stimulate Immunity

When a vaccine is administered, the immune system recognizes the introduced antigens as foreign. **Antigen-presenting cells (APCs)**, such as dendritic cells, engulf the vaccine components and present the antigens on their surface using **major histocompatibility complex (MHC)** molecules. This activates **helper T cells**, which coordinate the immune response.

Helper T cells stimulate **B cells** to produce specific antibodies against the antigen. These antibodies neutralize the pathogen and tag it for destruction by other immune cells. At the same time, **cytotoxic T cells** are activated to target and destroy infected cells. Importantly, the immune system generates **memory cells**—long-lived B and T cells that remain in the body for years, ready to mount a faster and stronger response if the real pathogen is encountered.

Herd Immunity and Public Health

Vaccines benefit not only individuals but also communities through **herd immunity**. When a significant portion of a population is vaccinated, the spread of a pathogen is significantly reduced, protecting those who cannot be vaccinated, such as newborns or individuals with medical conditions. For diseases like measles, which are highly contagious, achieving herd immunity requires over 90% of the population to be vaccinated.

Vaccination campaigns have eradicated or controlled many deadly diseases. **Smallpox**, which killed millions, was eradicated in 1980 through a global vaccination effort. Polio, once widespread, is now confined to a few regions due to intensive immunization programs. Routine childhood vaccinations prevent severe diseases like diphtheria, whooping cough, and tetanus, saving millions of lives annually.

Challenges and Innovations

Despite their success, vaccines face challenges such as vaccine hesitancy, access disparities, and the emergence of variants. Some pathogens, like HIV, evade vaccines due to their high mutation rates. However, advances in vaccine technology, including mRNA platforms and nanoparticle delivery systems, offer hope for tackling these challenges.

Vaccines remain one of the most effective tools for preventing infectious diseases, reducing mortality, and improving global health. Their impact extends beyond individuals, shaping entire populations and protecting future generations.

Disorders of the Immune System: Autoimmunity and Allergies

The immune system is designed to protect the body, but when its regulation fails, it can cause harm. **Autoimmune disorders** arise when the immune system mistakenly attacks the body's own tissues, while **allergies** result from an exaggerated immune response to harmless substances. Both conditions illustrate the complexity and precision required for proper immune function.

Autoimmunity

Autoimmune diseases occur when the immune system cannot distinguish between self and non-self. This leads to chronic inflammation and tissue damage as the immune system targets healthy cells.

1. **Mechanisms of Autoimmunity:**

   - **Loss of Self-Tolerance:** Normally, the immune system eliminates self-reactive T and B cells during their development in the thymus and bone marrow. When this process fails, self-reactive cells can persist and attack the body.
   - **Molecular Mimicry:** Some infections trigger autoimmunity when pathogen antigens resemble host proteins. For example, streptococcal infections can lead to **rheumatic fever**, where antibodies against the bacteria cross-react with heart tissue.
   - **Genetic and Environmental Factors:** Certain genes, such as those encoding MHC molecules, increase susceptibility to autoimmunity. Environmental triggers, including infections, toxins, and stress, can also contribute.

2. **Examples of Autoimmune Disorders:**

   - **Type 1 Diabetes:** The immune system destroys insulin-producing beta cells in the pancreas, leading to high blood sugar and dependence on insulin therapy.
   - **Rheumatoid Arthritis:** The immune system attacks joint tissues, causing inflammation, pain, and deformity.
   - **Systemic Lupus Erythematosus (SLE):** A systemic disease in which autoantibodies target DNA and other nuclear components, causing damage to multiple organs.
   - **Multiple Sclerosis:** The immune system damages the myelin sheath of nerve fibers, impairing nerve conduction and causing neurological symptoms.

Autoimmune diseases are often chronic and require treatments like immunosuppressants, corticosteroids, and biologic therapies to reduce immune activity and inflammation.

Allergies

Allergies result from an overactive immune response to harmless substances, known as allergens. Common allergens include pollen, dust mites, certain foods, and insect venom. The immune system perceives these substances as threats, triggering inflammation and other symptoms.

1. **Mechanisms of Allergic Reactions:**

    o   Allergies are mediated by **IgE antibodies**, which bind to allergens and activate **mast cells** and **basophils**. These cells release histamine and other inflammatory chemicals, causing symptoms like itching, swelling, and mucus production.
    o   The first exposure to an allergen sensitizes the immune system, leading to excessive IgE production. Subsequent exposures result in rapid and amplified allergic responses.

2. **Types of Allergic Reactions:**

    o   **Seasonal Allergies:** Pollen from plants triggers hay fever, characterized by sneezing, nasal congestion, and itchy eyes.
    o   **Food Allergies:** Common triggers include nuts, shellfish, and dairy. Symptoms range from mild (hives) to severe (anaphylaxis), a life-threatening reaction causing airway constriction and shock.
    o   **Asthma:** Allergens, respiratory infections, or irritants can cause airway inflammation and narrowing, leading to wheezing and difficulty breathing.
    o   **Contact Dermatitis:** Direct contact with allergens like poison ivy or nickel causes skin inflammation and itching.

3. **Treatment and Management:**

    o   **Antihistamines** block histamine receptors, reducing symptoms like itching and swelling.
    o   **Corticosteroids** control inflammation in conditions like asthma and allergic rhinitis.
    o   **Immunotherapy** (allergy shots) gradually desensitizes the immune system to specific allergens.
    o   **Epinephrine** is used in emergency cases of anaphylaxis to quickly reverse symptoms.

Understanding the mechanisms of autoimmunity and allergies highlights the delicate balance required for immune regulation. Treatments for these conditions aim to restore this balance, protecting the body while minimizing harm.

CHAPTER 15: THE REPRODUCTIVE SYSTEM

Anatomy of Male and Female Reproductive Systems

The reproductive system enables the production of offspring and ensures the continuation of the species. While the male system focuses on the production and delivery of sperm, the female system is designed for egg production, fertilization, and the development of offspring. Both systems include specialized organs and structures that work together to accomplish these functions.

Male Reproductive System

The male reproductive system consists of external and internal organs responsible for producing, storing, and delivering sperm.

External Structures:

1. **Penis:** The penis delivers sperm into the female reproductive tract during sexual intercourse. It consists of three columns of erectile tissue: two **corpora cavernosa** and one **corpus spongiosum**, which surrounds the urethra. During arousal, increased blood flow causes the penis to become erect, enabling penetration.
2. **Scrotum:** The scrotum is a pouch of skin and muscle that houses the **testes**. It regulates the temperature of the testes, keeping them slightly cooler than body temperature—an essential condition for healthy sperm production. The **cremaster muscle** contracts or relaxes to adjust the position of the testes relative to the body, aiding in temperature regulation.

Internal Structures:

1. **Testes:** The testes are paired oval glands located within the scrotum. They produce **sperm** and the hormone **testosterone**, which regulates secondary sexual characteristics such as increased muscle mass and deeper voice. Inside the testes are **seminiferous tubules**, where sperm are produced through a process called **spermatogenesis**.
2. **Epididymis:** The epididymis is a coiled tube located on the back of each testis. It stores and matures sperm, making them motile and capable of fertilization.
3. **Vas Deferens:** The vas deferens is a muscular tube that transports sperm from the epididymis to the **ejaculatory duct** during ejaculation. It passes through the **spermatic cord**, which also contains blood vessels and nerves.

4. **Seminal Vesicles:** These glands secrete a fluid rich in fructose, which provides energy for sperm. This fluid makes up a significant portion of semen.
5. **Prostate Gland:** The prostate surrounds the urethra and produces a milky fluid that enhances sperm motility and viability. This fluid also neutralizes the acidic environment of the female reproductive tract.
6. **Bulbourethral Glands:** Also known as Cowper's glands, these glands secrete a pre-ejaculatory fluid that lubricates the urethra and neutralizes traces of acidic urine.

Sperm Pathway: Sperm are produced in the seminiferous tubules, stored and matured in the epididymis, and transported via the vas deferens to the ejaculatory duct. During ejaculation, sperm mix with fluids from the seminal vesicles, prostate, and bulbourethral glands to form **semen**, which exits the body through the urethra.

Female Reproductive System

The female reproductive system is designed for egg production, fertilization, and the nurturing of a developing fetus. It includes external and internal structures, each with distinct functions.

External Structures (Vulva):

1. **Labia Majora and Labia Minora:** These folds of skin protect the external openings of the vagina and urethra. The **labia majora** contain fat and sweat glands, while the **labia minora** are thinner and more delicate.
2. **Clitoris:** The clitoris is a highly sensitive structure composed of erectile tissue. It is analogous to the male penis and becomes engorged during sexual arousal.
3. **Vestibule:** The vestibule is the area enclosed by the labia minora. It contains the **vaginal opening** and the **urethral opening**.
4. **Bartholin's Glands:** These glands produce mucus to lubricate the vaginal opening during sexual arousal.

Internal Structures:

1. **Ovaries:** The ovaries are paired almond-shaped glands located on either side of the uterus. They produce **oocytes (eggs)** and secrete the hormones **estrogen** and **progesterone**, which regulate the menstrual cycle and secondary sexual characteristics.
2. **Fallopian Tubes (Oviducts):** The fallopian tubes connect the ovaries to the uterus. They are the site of **fertilization**, where sperm meet the egg. The tubes contain tiny hair-like structures called **cilia**, which help guide the egg toward the uterus.
3. **Uterus:** The uterus is a hollow, muscular organ where a fertilized egg implants and develops into a fetus. Its walls have three layers:

- o **Endometrium:** The inner lining, which thickens during the menstrual cycle to support a potential pregnancy. If no pregnancy occurs, it sheds during menstruation.
- o **Myometrium:** The middle layer, made of smooth muscle, which contracts during childbirth.
- o **Perimetrium:** The outer protective layer.
4. **Cervix:** The cervix is the lower, narrow part of the uterus that opens into the vagina. It produces mucus that changes consistency during the menstrual cycle, facilitating or blocking sperm entry.
5. **Vagina:** The vagina is a muscular, elastic canal that serves as the birth canal and the receptacle for the penis during sexual intercourse. Its acidic environment helps prevent infections.

Ovarian and Menstrual Cycles: The female reproductive system operates on two interrelated cycles:

- **Ovarian Cycle:** This cycle governs the maturation and release of an egg. It consists of three phases:
  - o **Follicular Phase:** Follicles in the ovary mature under the influence of **follicle-stimulating hormone (FSH)**.
  - o **Ovulation:** Around day 14 of a typical 28-day cycle, a mature follicle releases an egg, triggered by a surge of **luteinizing hormone (LH)**.
  - o **Luteal Phase:** The ruptured follicle transforms into the **corpus luteum**, which secretes progesterone to prepare the uterus for implantation.
- **Menstrual Cycle:** The uterine lining thickens during the proliferative phase under the influence of estrogen, followed by the secretory phase driven by progesterone. If fertilization does not occur, hormone levels drop, leading to the shedding of the endometrium during menstruation.

Fertilization and Pregnancy: Fertilization typically occurs in the fallopian tube when a sperm penetrates an egg. The resulting **zygote** undergoes cell division as it travels to the uterus, where it implants into the endometrium. Hormones like progesterone and **human chorionic gonadotropin (hCG)** maintain the pregnancy, supporting the development of the placenta and fetus.

Hormonal Control of Reproduction

Reproduction is regulated by hormones that coordinate the development, maturation, and function of the reproductive organs. These hormones operate in tightly controlled feedback loops involving the hypothalamus, pituitary gland, and gonads. The hormonal control differs between males and females but ensures the production of gametes, regulation of sexual cycles, and preparation for reproduction.

Hormonal Regulation in Males

In males, the **hypothalamus** releases **gonadotropin-releasing hormone (GnRH)**, which stimulates the anterior pituitary to produce **luteinizing hormone (LH)** and **follicle-stimulating hormone (FSH)**. LH targets the **Leydig cells** in the testes, prompting them to produce **testosterone**, the primary male sex hormone. Testosterone supports **spermatogenesis**, the production of sperm in the **seminiferous tubules**, and maintains secondary sexual characteristics like increased muscle mass and facial hair.

FSH works alongside testosterone to stimulate the **Sertoli cells**, which provide nourishment and structural support for developing sperm. Sertoli cells also release **inhibin**, a hormone that negatively regulates FSH secretion, ensuring that sperm production remains balanced. Testosterone itself exerts a negative feedback effect on the hypothalamus and pituitary, preventing overproduction of LH and maintaining hormonal stability.

Hormonal Regulation in Females

In females, the hypothalamus also releases GnRH, triggering the anterior pituitary to secrete LH and FSH. These hormones control the **ovarian cycle** and are responsible for follicular development, ovulation, and preparation of the uterus for pregnancy.

FSH stimulates the maturation of ovarian follicles, which secrete **estrogen**. Estrogen has dual roles: at low levels, it inhibits FSH and LH production (negative feedback), but as its levels peak during the late follicular phase, it stimulates a surge in LH (positive feedback). This LH surge triggers **ovulation**, the release of a mature egg from the ovary.

After ovulation, the ruptured follicle forms the **corpus luteum**, which secretes **progesterone** and smaller amounts of estrogen. Progesterone prepares the uterine lining for potential implantation and suppresses further release of GnRH, LH, and FSH to prevent the maturation of additional follicles.

If fertilization does not occur, the corpus luteum degenerates, progesterone levels drop, and the **endometrium** is shed during menstruation. If fertilization happens, **human chorionic gonadotropin (hCG)**, secreted by the developing embryo, maintains the corpus luteum until the placenta takes over hormone production.

The Biology of Fertilization and Pregnancy

Fertilization and pregnancy involve complex biological processes that begin with the union of sperm and egg and culminate in the development of a fetus within the uterus. These events require precise coordination of gametes, hormones, and cellular interactions.

Fertilization

Fertilization occurs in the **fallopian tube**, typically within 12-24 hours after ovulation. For fertilization to take place, sperm must travel through the female reproductive tract, navigating the **cervix**, uterus, and into the fallopian tube. This journey is aided by uterine contractions and the presence of cervical mucus, which becomes thinner and more hospitable around ovulation.

Once sperm reach the egg, they undergo **capacitation**, a biochemical process that enhances their ability to penetrate the egg's protective layers. The egg is surrounded by the **zona pellucida**, a glycoprotein shell, and the **corona radiata**, a cluster of supportive cells. A sperm binds to receptors on the zona pellucida, triggering the **acrosome reaction**, where enzymes are released to digest the zona and allow the sperm to reach the egg's plasma membrane.

When a sperm successfully fuses with the egg, its nucleus is injected into the egg's cytoplasm, and the zona pellucida undergoes a **cortical reaction** to prevent other sperm from entering. The sperm and egg nuclei combine, forming a **zygote** with a complete set of chromosomes (46 in humans).

Early Development and Implantation

The zygote undergoes rapid cell divisions, forming a **morula** and then a **blastocyst**. The blastocyst reaches the uterus about five days after fertilization and implants into the **endometrium**. The **trophoblast** cells of the blastocyst invade the uterine lining, forming the **placenta**, which establishes nutrient and waste exchange between mother and fetus.

Hormonal Support of Pregnancy

During early pregnancy, hCG maintains the corpus luteum, ensuring the continued production of progesterone and estrogen to sustain the endometrium. By the second trimester, the **placenta** takes over hormone production. Progesterone maintains uterine quiescence, preventing contractions, while estrogen promotes uterine growth and enhances blood flow to the placenta.

Fetal Development

Over the course of 40 weeks, the fetus grows from a few cells to a fully developed baby. By the end of the first trimester, organ systems begin to form (organogenesis). The second trimester is marked by rapid growth and functional development, while the third trimester prepares the fetus for birth, with final maturation of the lungs and nervous system.

Reproductive Technologies: Innovations and Ethical Considerations

Reproductive technologies have advanced significantly, providing solutions for infertility and enabling new possibilities for conception. These innovations range from **assisted reproductive technologies (ART)** like in vitro fertilization (IVF) to emerging techniques like gene editing. While these technologies offer hope, they also raise ethical and social concerns.

Assisted Reproductive Technologies

1. **In Vitro Fertilization (IVF):** IVF involves retrieving eggs from a woman's ovaries, fertilizing them with sperm in a laboratory, and transferring the resulting embryos into the uterus. This method is commonly used for infertility due to blocked fallopian tubes, low sperm count, or unexplained infertility. IVF often includes the use of **hormonal stimulation** to produce multiple eggs and increases the chances of success.

2. **Intracytoplasmic Sperm Injection (ICSI):** In ICSI, a single sperm is directly injected into an egg to overcome severe male infertility. This technique ensures fertilization even with very low sperm count or motility.

3. **Egg and Sperm Donation:** Donor eggs or sperm allow individuals who cannot produce viable gametes to conceive. This is particularly helpful for older women, individuals with genetic disorders, or same-sex couples.

4. **Surrogacy:** Surrogacy involves another individual carrying a pregnancy for intended parents. It can be gestational (where the surrogate carries an embryo created using IVF) or traditional (where the surrogate's egg is used).

Emerging Technologies

1. **Gene Editing:** Techniques like **CRISPR-Cas9** enable precise modification of DNA in embryos. While this has potential for preventing genetic diseases, it raises concerns about unintended effects and the ethical implications of altering the human genome.

2. **Three-Parent IVF:** This technique replaces defective mitochondrial DNA in an egg with healthy mitochondrial DNA from a donor, preventing mitochondrial diseases. However, it introduces genetic material from three individuals, sparking debates about identity and inheritance.

3. **Artificial Gametes:** Scientists are exploring the creation of sperm and eggs from stem cells, which could allow individuals with no viable gametes to have biological children.

Ethical Considerations

While reproductive technologies offer hope, they pose challenges:

- **Access and Equity:** High costs limit availability, creating disparities in who can benefit.
- **Embryo Selection and Disposal:** Decisions about which embryos to implant or discard raise moral questions, especially regarding unused embryos.
- **Gene Editing and "Designer Babies":** Editing genes for non-medical traits, such as intelligence or physical appearance, could lead to societal inequality and loss of genetic diversity.
- **Surrogacy Exploitation:** In some cases, surrogates, particularly in low-income regions, may face exploitation or inadequate legal protections.

Reproductive technologies continue to push the boundaries of biology and ethics, transforming how society approaches fertility and family building. Balancing innovation with responsible use is essential as these technologies evolve.

CHAPTER 16: INTEGRATION AND INTERDEPENDENCE IN HUMAN BIOLOGY

The Relationship Between Body Systems in Maintaining Homeostasis

Homeostasis is the body's ability to maintain a stable internal environment despite external changes. This balance is achieved through the integration and interdependence of various body systems. Each system performs specific tasks but works closely with others to regulate variables such as temperature, pH, blood pressure, and nutrient levels. Disruptions in one system often affect others, demonstrating the complexity and coordination required to sustain life.

Nervous and Endocrine Systems: Master Regulators

The **nervous system** and **endocrine system** coordinate most homeostatic processes by detecting changes in the internal environment and responding with precise adjustments. The nervous system uses electrical signals for rapid responses, while the endocrine system employs hormones for slower, longer-lasting effects.

The **hypothalamus**, part of the nervous system, acts as the body's control center. It monitors variables like temperature, osmolarity, and blood pressure, sending signals to initiate corrective actions. For example, when body temperature rises, the hypothalamus signals sweat glands to release sweat and dilates blood vessels in the skin, promoting heat loss. The endocrine system complements this response by releasing hormones like **thyroxine**, which adjusts metabolic rates to produce or conserve heat.

In regulating blood glucose, the nervous and endocrine systems collaborate. When blood sugar levels rise after a meal, the pancreas (an endocrine organ) secretes **insulin**, prompting cells to absorb glucose. If levels drop too low, the pancreas releases **glucagon**, stimulating the liver to release stored glucose. The autonomic nervous system can amplify these effects during stress, ensuring the body has adequate energy.

Cardiovascular and Respiratory Systems: Oxygen and pH Balance

The **cardiovascular system** and **respiratory system** work together to deliver oxygen to tissues and remove carbon dioxide, a waste product of cellular respiration. This partnership also maintains blood pH, which is tightly regulated between 7.35 and 7.45.

When CO_2 levels in the blood rise, it reacts with water to form carbonic acid, lowering pH. Chemoreceptors in the brainstem detect this change and increase the respiratory rate. Faster breathing expels more CO_2, reducing acidity and restoring pH balance. At the same time, the cardiovascular system increases heart rate and blood flow to ensure efficient oxygen delivery and CO_2 removal.

The **renal system** supports these processes by excreting hydrogen ions and reabsorbing bicarbonate, providing long-term pH regulation. If the lungs or kidneys fail to compensate, acidosis or alkalosis can occur, disrupting enzyme activity and cellular functions.

Digestive and Circulatory Systems: Nutrient Distribution

The **digestive system** breaks down food into nutrients like glucose, amino acids, and fatty acids, which are absorbed into the bloodstream. The **circulatory system** then distributes these nutrients to cells throughout the body.

The liver acts as an intermediary between these systems, processing nutrients absorbed from the intestines. For example, the liver converts excess glucose into glycogen for storage and releases it during fasting to maintain blood sugar levels. It also detoxifies harmful substances, ensuring that the blood circulating to tissues remains safe.

After a meal, the digestive and circulatory systems interact with the endocrine system to regulate nutrient availability. **Insulin** helps cells absorb glucose, while hormones like **cholecystokinin (CCK)** and **ghrelin** signal fullness or hunger, influencing when and how much we eat.

Musculoskeletal and Nervous Systems: Movement and Protection

The **musculoskeletal system** provides structure, supports movement, and protects vital organs, but its function depends heavily on the **nervous system**. Nerves stimulate skeletal muscles to contract, enabling voluntary movements like walking or lifting objects. Reflex arcs, such as the withdrawal reflex, illustrate how these systems work together to protect the body from harm. When you touch something hot, sensory nerves send a signal to the spinal cord, which immediately activates motor neurons to pull your hand away.

Bones also support homeostasis by storing minerals like calcium and phosphate. When blood calcium levels drop, the **parathyroid glands** release **parathyroid hormone (PTH)**, which signals bones to release calcium into the bloodstream. The **renal system** and digestive system contribute by adjusting calcium excretion and absorption, ensuring that muscle contractions, nerve signaling, and bone strength remain optimal.

Immune and Lymphatic Systems: Defense and Fluid Balance

The **immune system** defends against pathogens, while the **lymphatic system** maintains fluid balance and facilitates immune responses. Lymphatic vessels collect excess interstitial fluid and return it to the circulatory system, preventing tissue swelling (edema). Along the way, lymph passes through **lymph nodes**, where immune cells like lymphocytes detect and neutralize pathogens.

The circulatory system supports these defenses by transporting white blood cells to infection sites. Inflammation, a key immune response, involves increased blood flow to damaged tissues, delivering immune cells and nutrients necessary for repair. The **endocrine system** regulates this response with hormones like **cortisol**, which dampens excessive inflammation to prevent tissue damage.

Urinary and Cardiovascular Systems: Fluid and Electrolyte Balance

The **urinary system** and **cardiovascular system** maintain fluid and electrolyte homeostasis by regulating blood volume and composition. The kidneys filter blood, removing waste products like urea while adjusting water and salt levels to maintain blood pressure.

When blood volume drops, the kidneys release **renin**, activating the **renin-angiotensin-aldosterone system (RAAS)**. Angiotensin II constricts blood vessels to raise blood pressure, while aldosterone promotes sodium and water reabsorption in the kidneys. Conversely, when blood volume is too high, the heart secretes **atrial natriuretic peptide (ANP)**, which reduces sodium reabsorption and promotes urine production.

The cardiovascular and urinary systems also collaborate to remove metabolic waste. For example, the kidneys excrete excess hydrogen ions to regulate blood pH, while the heart and blood vessels ensure consistent delivery of blood to the kidneys for filtration.

Reproductive and Endocrine Systems: Hormonal Control

The **reproductive system** relies on hormones from the **endocrine system** for development, function, and regulation. In females, the hypothalamus and pituitary gland release **GnRH**, **FSH**, and **LH**, which control the ovarian cycle and menstrual cycle. In males, these hormones regulate testosterone production and spermatogenesis.

During pregnancy, the reproductive and endocrine systems work with the cardiovascular system to support the developing fetus. The placenta produces hormones like **hCG** and progesterone, which maintain the uterine lining and prevent contractions. The cardiovascular system increases blood volume to meet the oxygen and nutrient demands of the growing fetus.

Integumentary and Nervous Systems: Temperature Regulation

The **integumentary system** (skin, hair, and nails) and **nervous system** regulate body temperature. When body temperature rises, the hypothalamus signals sweat glands to release sweat, which evaporates and cools the skin. Blood vessels in the skin dilate, increasing heat loss through radiation.

In cold conditions, the hypothalamus triggers **vasoconstriction**, reducing blood flow to the skin to conserve heat. Shivering, caused by rapid muscle contractions, generates additional heat to maintain core temperature. These responses demonstrate the close integration of multiple systems in thermoregulation.

How Stress Affects Biological Systems

Stress triggers a cascade of physiological responses that affect multiple biological systems, highlighting their interdependence in managing and mitigating the impact. The **nervous system**, particularly the hypothalamus, initiates the stress response by activating the **sympathetic nervous system** and the **hypothalamic-pituitary-adrenal (HPA) axis**. These mechanisms prepare the body for immediate survival, but prolonged activation can disrupt homeostasis and harm various systems.

When stress is perceived, the hypothalamus signals the adrenal glands to release **adrenaline** and **noradrenaline**. These hormones increase heart rate, blood pressure, and respiratory rate, ensuring that oxygen and nutrients reach vital organs quickly. The **cardiovascular system** supports this fight-or-flight response, but chronic stress strains the heart and vessels, increasing the risk of hypertension, atherosclerosis, and heart disease.

Simultaneously, the HPA axis promotes the release of **cortisol**, a glucocorticoid that regulates energy by increasing glucose availability. Cortisol stimulates the **liver** to release glucose into the bloodstream, providing immediate fuel for muscles and the brain. However, prolonged cortisol elevation can impair **insulin sensitivity**, leading to metabolic conditions like type 2 diabetes. The **digestive system** is also affected, with stress slowing gastric emptying and altering gut motility, which can result in symptoms like nausea, diarrhea, or constipation.

Stress significantly impacts the **immune system**, initially enhancing its activity but later suppressing it. Acute stress stimulates **natural killer (NK) cells** and pro-inflammatory cytokines, bolstering the immune defense temporarily. Chronic stress, however, suppresses immune function, reducing the production of lymphocytes and leaving the body vulnerable to infections and slower wound healing.

The **reproductive system** is another target. In females, stress can disrupt the menstrual cycle by suppressing gonadotropin-releasing hormone (GnRH) and reducing estrogen and progesterone production. In males, stress lowers testosterone levels and can impair spermatogenesis. Both genders may experience reduced libido due to the physiological and psychological impacts of stress.

The **nervous system** itself experiences wear from chronic stress. Elevated cortisol can damage the hippocampus, affecting memory and learning. Stress also exacerbates mental health conditions, such as anxiety and depression, by altering neurotransmitter balance and reducing neurogenesis in key brain regions.

Advances in Artificial Organs and Biotechnology

Artificial organs and biotechnology are reshaping how human biology integrates with technology to address organ failure and enhance quality of life. These advancements demonstrate the interplay between biological systems and engineered solutions, offering new ways to restore or replace lost function.

Artificial Hearts have transformed the treatment of end-stage heart failure. Devices like the **total artificial heart (TAH)** completely replace the function of the ventricles, ensuring continuous blood circulation. While initially used as a bridge to transplantation, modern artificial hearts, such as the SynCardia TAH, are increasingly being considered as long-term solutions. These systems require external power sources and pumps, creating challenges in patient mobility and quality of life, but ongoing innovations aim to improve portability and integration.

Kidney dialysis was one of the earliest artificial organ technologies, and progress continues with wearable and implantable dialysis devices. Emerging technologies include **bioartificial kidneys**, which combine synthetic filters with live kidney cells to mimic natural filtration and endocrine functions. These devices aim to eliminate the need for traditional dialysis sessions by restoring kidney function within the body.

3D bioprinting has enabled the creation of customized tissues and organs. Using patient-derived stem cells, bioprinters can construct structures like **tracheas**, **bladders**, and even sections of **livers** for implantation. This approach reduces the risk of immune rejection, as the printed tissues are genetically identical to the recipient's own cells. Researchers are also exploring bioprinted **vascularized tissues**, which address the challenge of creating blood supply networks within artificial organs.

The integration of **neuroprosthetics** with the nervous system exemplifies the synergy between biology and biotechnology. Devices like **cochlear implants** restore hearing by directly stimulating auditory nerves, while **retinal implants** partially restore vision in people with degenerative eye diseases. Advanced prosthetics now incorporate **brain-machine interfaces**, allowing users to control artificial limbs using neural signals, effectively blending the musculoskeletal and nervous systems with engineered solutions.

CRISPR-Cas9 gene-editing technology has opened possibilities for engineering tissues resistant to diseases or modifying donor organs to prevent rejection. For example, gene-edited pig organs are being developed for **xenotransplantation**, potentially solving the shortage of human donor organs. Ethical concerns about editing human embryos and long-term effects on ecosystems and populations remain under debate.

Biotechnology is also advancing **artificial skin**, which integrates sensory feedback systems to mimic touch and temperature perception. These innovations are particularly transformative for burn victims and those with sensory deficits, as they restore both protective and communicative functions of the **integumentary system**.

The fusion of biological systems and artificial devices is moving toward complete integration, creating a future where engineered solutions enhance natural functions and support survival in conditions previously deemed untreatable.

Future Directions in Human Biology Research

Human biology research continues to expand, focusing on understanding complex systems, addressing diseases, and exploring human adaptability. Advances in genomics, regenerative medicine, and computational modeling are shaping future directions, emphasizing the integration of biology, technology, and ethics.

Personalized Medicine is a key area of focus. By leveraging insights from the **human genome project**, researchers are tailoring treatments based on individual genetic profiles. For example, pharmacogenomics is optimizing drug therapies to improve efficacy and reduce side effects. This approach relies on understanding how genetic variations influence drug metabolism, making treatments more precise for conditions like cancer and autoimmune diseases.

Regenerative medicine is advancing with the use of stem cells to repair damaged tissues and organs. **Induced pluripotent stem cells (iPSCs)**, which are reprogrammed from adult cells, offer the potential to regenerate tissues without ethical concerns associated with embryonic stem cells. These technologies aim to treat conditions like spinal cord injuries, degenerative diseases, and organ failure by promoting natural healing and reducing reliance on donor transplants.

Epigenetics is revealing how environmental factors influence gene expression without altering DNA sequences. This field is shedding light on how diet, stress, and toxins affect long-term health and the development of diseases like diabetes and cancer. Understanding epigenetic mechanisms could lead to new therapies that reverse harmful gene expression patterns.

The **microbiome** is another frontier in human biology research. Scientists are uncovering how gut bacteria interact with the immune, digestive, and nervous systems to influence health. Probiotic and prebiotic therapies are being explored to treat conditions like inflammatory bowel disease, obesity, and even mental health disorders, emphasizing the microbiome's systemic impact.

Bioinformatics and **AI-driven modeling** are enabling researchers to simulate complex biological processes. By analyzing vast datasets, AI can identify patterns in disease progression, predict drug interactions, and propose new treatment pathways. Computational models are also being used to simulate entire organ systems, accelerating research on conditions like heart disease and kidney failure.

Research into **aging** is uncovering mechanisms that could extend healthy lifespans. Studies on **telomeres**, cellular senescence, and mitochondrial function aim to slow or reverse the biological processes that contribute to age-related diseases. Experimental drugs like **senolytics**, which clear out aging cells, are being tested to improve healthspan without compromising longevity.

Human biology is also exploring **space biology**, studying how the body adapts to microgravity and radiation during long-term space travel. Insights into muscle atrophy, bone density loss, and fluid shifts in astronauts are informing strategies for maintaining health during missions to Mars and beyond. These findings have implications for understanding similar conditions on Earth, such as osteoporosis and cardiovascular diseases.

Ethical considerations are becoming increasingly important as research pushes boundaries. Advances in **gene editing**, **cloning**, and human enhancement raise questions about safety, fairness, and societal impact. Ensuring equitable access to new technologies and avoiding misuse are priorities as science reshapes what it means to be human.

Future research in human biology is converging on a deeper understanding of integration and adaptability, emphasizing the dynamic interplay between biology and technology. By addressing challenges across scales, from molecular mechanisms to planetary survival, human biology aims to enhance health, resilience, and the potential for innovation.

APPENDIX

Terms and Definitions

- **Homeostasis**: The process by which the body maintains a stable internal environment despite external changes.
- **Neuron**: A nerve cell that transmits electrical signals in the nervous system.
- **Synapse**: The junction between two neurons where neurotransmitters facilitate communication.
- **Neurotransmitter**: Chemical messenger that transmits signals across synapses between neurons.
- **Axon**: A long, thread-like part of a neuron that transmits signals away from the cell body.
- **Dendrite**: Branched extensions of a neuron that receive signals from other cells.
- **Glial Cells**: Supportive cells in the nervous system that protect and nourish neurons.
- **Myelin Sheath**: A fatty layer surrounding axons that speeds up signal transmission.
- **Central Nervous System (CNS)**: The brain and spinal cord, responsible for processing information.
- **Peripheral Nervous System (PNS)**: Nerves outside the CNS that transmit signals to and from the body.
- **Endocrine System**: A system of glands that produce hormones to regulate body functions.
- **Hormone**: A chemical messenger secreted by glands into the bloodstream.
- **Thyroid Gland**: Produces hormones that regulate metabolism and energy use.
- **Pituitary Gland**: Often called the "master gland," it controls other endocrine glands.
- **Insulin**: A hormone produced by the pancreas that regulates blood glucose levels.
- **Glucagon**: A hormone that raises blood glucose levels by stimulating glycogen breakdown.
- **Adrenal Glands**: Glands above the kidneys that produce adrenaline and cortisol.
- **Testosterone**: A male sex hormone that supports sperm production and secondary sexual characteristics.
- **Estrogen**: A female sex hormone responsible for regulating the menstrual cycle and secondary sexual characteristics.
- **Ovary**: The female reproductive organ that produces eggs and hormones like estrogen and progesterone.
- **Testes**: Male reproductive organs that produce sperm and testosterone.
- **Zygote**: A fertilized egg cell formed by the union of sperm and egg.
- **Embryo**: The early stage of development in multicellular organisms following fertilization.

- **Fetus**: A developing human from the end of the embryonic stage (8 weeks) until birth.
- **Placenta**: An organ that provides oxygen and nutrients to the fetus while removing waste products.
- **Amniotic Sac**: A fluid-filled sac that cushions and protects the developing fetus.
- **Oogenesis**: The process of egg cell production in the ovaries.
- **Mitochondria**: Organelles in cells responsible for energy production.
- **Ribosome**: A cellular structure that synthesizes proteins.
- **Nucleus**: The control center of a cell, containing DNA.
- **DNA (Deoxyribonucleic Acid)**: A molecule that carries genetic instructions for growth and function.
- **RNA (Ribonucleic Acid)**: A molecule involved in protein synthesis and gene regulation.
- **Chromosome**: A structure of DNA and protein that carries genetic information.
- **Gene**: A segment of DNA that codes for a specific protein.
- **Genome**: The complete set of an organism's genetic material.
- **Mutation**: A change in the DNA sequence that can affect genetic expression.
- **Epigenetics**: The study of changes in gene expression without altering the DNA sequence.
- **Cell Membrane**: The outer boundary of a cell that regulates what enters and exits.
- **Cytoplasm**: The gel-like substance inside a cell that contains organelles.
- **Lysosome**: A cellular organelle containing enzymes for breaking down waste.
- **Golgi Apparatus**: An organelle involved in modifying, sorting, and packaging proteins.
- **Endoplasmic Reticulum (ER)**: A network of membranes for protein and lipid synthesis.
- **ATP (Adenosine Triphosphate)**: The primary molecule for energy transfer in cells.
- **Osmosis**: The movement of water across a semipermeable membrane.
- **Diffusion**: The movement of molecules from an area of higher concentration to lower concentration.
- **Enzyme**: A protein that speeds up chemical reactions in the body.
- **Hemoglobin**: A protein in red blood cells that carries oxygen.
- **Plasma**: The liquid component of blood that carries cells and nutrients.
- **Platelet**: A cell fragment involved in blood clotting.
- **Lymphocyte**: A type of white blood cell important in immune responses.
- **Antibody**: A protein produced by B cells that targets specific antigens.
- **Antigen**: A substance that triggers an immune response.
- **Phagocyte**: A cell that engulfs and digests pathogens and debris.
- **Macrophage**: A large phagocytic cell involved in immune defense.
- **Inflammation**: The body's response to injury or infection, characterized by redness and swelling.
- **Autoimmune Disease**: A condition where the immune system attacks the body's own cells.
- **Pathogen**: A microorganism that causes disease.
- **Virus**: A microscopic infectious agent that replicates inside living cells.

- **Bacteria**: Single-celled organisms that can be beneficial or harmful.
- **Fungus**: A type of organism, including yeasts and molds, that can cause infections.
- **Bone Marrow**: The soft tissue inside bones where blood cells are produced.
- **Cartilage**: A flexible connective tissue found in joints and the respiratory tract.
- **Ligament**: A connective tissue that connects bones to other bones.
- **Tendon**: A connective tissue that attaches muscles to bones.
- **Joint**: A point where two or more bones meet, allowing movement.
- **Peristalsis**: Wave-like muscle contractions that move food through the digestive tract.
- **Esophagus**: A muscular tube connecting the throat to the stomach.
- **Stomach**: A digestive organ that breaks down food with acid and enzymes.
- **Small Intestine**: The organ where most nutrient absorption occurs.
- **Large Intestine**: The organ that absorbs water and forms feces.
- **Liver**: An organ that processes nutrients, detoxifies substances, and produces bile.
- **Pancreas**: An organ that produces digestive enzymes and hormones like insulin.
- **Kidney**: An organ that filters blood to produce urine.
- **Bladder**: A hollow organ that stores urine.
- **Nephron**: The functional unit of the kidney that filters blood.
- **Ureter**: A tube that carries urine from the kidneys to the bladder.
- **Diaphragm**: A muscle that aids in breathing by expanding and contracting the lungs.
- **Alveoli**: Tiny air sacs in the lungs where gas exchange occurs.
- **Bronchi**: Airways that branch from the trachea into the lungs.
- **Trachea**: The windpipe that connects the larynx to the bronchi.
- **Larynx**: The voice box, involved in sound production and breathing.
- **Epiglottis**: A flap that prevents food from entering the trachea during swallowing.
- **Artery**: A blood vessel that carries oxygen-rich blood away from the heart.
- **Vein**: A blood vessel that carries oxygen-depleted blood back to the heart.
- **Capillary**: A tiny blood vessel where exchange of gases and nutrients occurs.
- **Aorta**: The largest artery in the body.
- **Pulmonary Circulation**: The part of the circulatory system that moves blood between the heart and lungs.
- **Systemic Circulation**: The part of the circulatory system that moves blood between the heart and the rest of the body.
- **Hypoxia**: A condition where tissues lack sufficient oxygen.
- **Hypertension**: High blood pressure, a condition that strains the cardiovascular system.
- **Anemia**: A condition characterized by a lack of healthy red blood cells.
- **Stroke**: A medical emergency caused by interrupted blood flow to the brain.
- **Diabetes**: A metabolic disorder characterized by high blood sugar levels.
- **Osteoporosis**: A condition where bones become weak and brittle.
- **Metabolism**: The set of chemical reactions that sustain life by converting food into energy.
- **Biotechnology**: The use of biological processes, organisms, or systems to develop products or technologies.

AFTERWORD

Congratulations on completing *Human Biology Step by Step*! I hope this overview has deepened your understanding of the human body and left you with a sense of awe at the complexity and brilliance of life.

As we explored each chapter, we uncovered how every component of the body—whether it's a single cell or an entire organ system—has a critical role in keeping us alive and thriving. From the delicate balance maintained by homeostasis to the intricate relationship between our nervous, cardiovascular, and immune systems, human biology is a testament to the power of cooperation and adaptation.

Science has given us remarkable tools to study and improve the human body, from decoding the mysteries of DNA to developing life-saving medical technologies. Yet, there's still so much we don't know. Every discovery raises new questions, pushing the boundaries of our understanding and challenging us to rethink what we thought we knew.

This book is just the beginning. Human biology is a dynamic field, constantly evolving with new research and innovations. As we look to the future, exciting advancements are on the horizon. From personalized medicine based on genetic profiles to artificial organs and cutting-edge therapies for once-incurable diseases, the possibilities are extraordinary.

But beyond the science, understanding human biology has a deeply personal dimension. It helps us take better care of our bodies, make informed health decisions, and appreciate the miracle of being alive. It reminds us that every heartbeat, every breath, every movement is part of a larger story—one that connects us to each other and to the vast web of life on Earth.

As you close this final chapter, remember that the learning doesn't stop here. The study of human biology isn't just about knowledge—it's about curiosity, discovery, and a lifelong appreciation for the incredible system that sustains us.

Whether you're a student, a professional, or simply someone curious about how the body works, I hope you've found this book to be a valuable resource and inspires you to keep asking questions, keep exploring, and keep marveling at the wonder that is human life.

Here's to the science of being human, and to the endless possibilities it holds.